# FREEDOM to MOVE

Cataloging-in-Publication Data is available from the Library of Congress.

978-1-62671-202-7 (hardback)
978-1-62671-135-8 (paperback)
978-1-62671-136-5 (epub)
978-1-62671-137-2 (epdf)

Cover image: People walking through the busy intersection at Fifth Avenue and Twenty-Third Street in New York City on a summer day with sunset flare between the background buildings: deberarr/iStock via Getty Images Plus.

# FREEDOM to MOVE

## Restoring Choice to America's Transportation

**Alan Cunningham**

Purdue University Press | West Lafayette, Indiana

# Contents

# Preface

I biked to the café where I'm writing this. I like writing anonymously in cafés, and I know every place within a three-mile radius of home that will let you loiter with the internet. During the winter I took my car almost every time. As a result, biking is a new ordeal every spring. No matter. Hills that were impossible a month ago are getting possible, but only if I keep biking them.

Biking in my neighborhood is an insurgent act, as it would be in most of America. By state law, I should be biking in the shoulder of the highway. The posted speed limit is 35 mph, but we all know the speed driven by automobiles in free-flowing traffic is 50 mph with a design speed of 45 mph. My chances of being killed if I'm hit at 35 mph are 50/50, but they are 75/25 if I'm hit at 50 mph. These speeds have only increased since COVID-19 lockdowns, work from home, and emptier streets. Knowing this, I bike on the sidewalk, but carefully, respecting walkers. The sidewalk has been the province of all non-automotive movement for my entire lifetime. For the last century, we have assumed that every place worth going to has to be reached by traffic, even when you could walk or bike to it instead.

> Throughout this book, the term "traffic" is used to refer to travel by car on our network of roads built to accommodate movement by motorized vehicles of all sizes. "Traffic" is compared with "walking," "biking," and "transit" later in the book, and a single two-syllable word was chosen for each of the four modes in this book about multimodalism.

While many people have misgivings about traffic for many reasons, traffic is too useful to casually throw away. But it is also too expensive, directly and in its side effects, to remain our sole mode of getting around. I wrote this book to answer the question: "How does America get off of traffic?" Even though the question has been answered many times, America remains steadfastly dependent on traffic and its trappings, so we need to look for more insightful answers.

Many in the sustainability community—such as transit advocates, bike advocates, walking advocates, land use advocates, equitable housing and transportation advocates, design advocates, wildlife advocates, air quality advocates, and water quality advocates—see traffic as evil, robbing us of human-scaled places, indebting us to oil-rich foreign powers, and robbing our children of a stable climate. If traffic is so bad, why does it persist? Are people who drive cars evil? This is the wrong question; thinking of any issue as evil blinds us to the needs that the "evil" is fulfilling. Traffic is not so much evil as it is too useful—beguiling, not sinister. Demonizing traffic is not a policy, it is an alibi for the absence of policy.[1] Books of the 1970s and '90s, such as *Superhighway—Superhoax* by Helen Leavitt and *The City After the Automobile* by Moshe Safdie, predicted the imminent demise of the car and of traffic over forty years ago. I've spent my whole life waiting for that demise. My parents and grandparents never knew an America without traffic as the dominant mode of transportation.

Traffic—with its advantages of great financing, convenience, speed, privacy, and range—can never be replaced. Traffic cannot be our only choice, as it is too expensive; forces us to buy, operate, and maintain cars that pollute our air, land, and water; and makes every other mode of transportation useless if not dangerous. America deserves choices, but right now all we have is traffic.

I started writing this after reading arguments for and against our dependence on traffic. I saw authors using case studies to bolster their arguments for and against our reliance on car travel—such as a perfectly designed high-end subdivision, a walkable mixed-use mall off the highway, or a failing transit station area—and saw that case studies could be picked to support whatever argument the author might choose. To address the question of getting past traffic dependence myself, I resolved to look at the conditions in America as a whole. I chose to look at quantitative transportation and land use data for the whole country at once, both with national averages and with the complete set of census block groups in the US. I also wanted to get an objective comparison of traffic with biking, walking, and rail transit. The results of this research are in your hands.

- The first couple of chapters review the history of America's national and local transportation and land use over the centuries and appreciate what we have today after our long history going slowly through the mud.

- The middle chapters explore our current conditions in detail, from issues such as traffic collisions and parking requirements to a comparison of characteristics, relative costs, and financing for walking, biking, traffic, and transit. How does traffic compare to rail transit in land use performance, using nationally applicable measures like land use intensity? This comparison shows that other modes of transport are better than traffic in many ways, but not in privacy, speed, or network extent. What could we do if transit affected land use as much as traffic?

- The concluding chapters propose human-scaled commuting, allowing transit station areas to work as hubs of development, to become origins and destinations of transit and active trips. The 1 percent of America within a bikeable distance of rail transit stations could be developed walkably and bikeably and hold 50 percent more jobs and housing than exist in America today. What can we achieve if we open more doors (offices, retail, residential) closer to each other and to transit, including gains in water quality, health, household wealth, and transportation budgets? Focusing development of high-value land use within this 1 percent of land can produce as much economic and people-centered activity within bikeable and walkable distances of transit as we currently achieve within comparable distances of roadways.

I have always been more interested in finding solutions than bemoaning problems; always more interested in the unifying goal rather than the divisive fight. This book steps outside the realm of design and advocacy and shows "Here is how America could get over traffic," and the rewards America could experience for a small change in policy. America needs a path forward that works from where we are, with the tools that we've got, not a fantasy that works from where we should be, with tools from overseas. This book shows how we can use what we've got to get to a safer, more affordable, more diverse, and more prosperous America. Traffic is a great but expensive tool, but we deserve to be able to choose safer or less expensive tools. This book proposes restoring choice to American transportation and land use.

# 1

## Stuck in Traffic

Over the last century, America has gone all in on traffic—the use of individually piloted motorized vehicles driven on exclusive trafficways within public rights of way. We have built a vast and functional road network since we started to pave our streets for cars instead of horses at the beginning of the twentieth century. Traffic gives us unprecedented choice, speed, mobility, and privacy. Why would anyone with the means to own a car not own one and drive it everywhere?

Automotive traffic served our needs better than the horse, streetcar, and dirt-mud road network we had in 1900. Horses ejected pounds and gallons of waste every day, were short-lived, and were liable to bite and kick during their days of toil. The streetcar was much less efficient than streetcars today, but still silent and deadly in streets full of horse-drawn wagons and people on foot. The streets of every town and city predated underground sewer drainage networks and carried every stream we now relegate to pipes deep underground. Traffic now kills tens of thousands of Americans and costs us tens of billions every year. It is past time to reconsider better alternatives.

The transportation we now have in America works pretty well provided that you have your own car, and you are not traveling in the same direction and at the same time as everyone else. The America we have today is built for cars, not people. More precisely, it is not built for anything but cars—not for walkers, bikers, or transit users. If you don't have a car, transportation works pretty badly for you in much of America. Everything is at the scale of the car. We have a century-old tradition of building and zoning for traffic, and we have only gotten better at it over the decades.

The land use we now have in America is mostly built out at a scale only usable by cars in traffic. Zoning codes and parking regulations developed at exactly the same time as automobiles in traffic, to serve the spatial needs of traffic. As with transportation, much of America is not built for anything but cars. Walkers and bikers have too far to go to reach anything useful, and the land use around transit stops is too sparsely developed for most routes to be serviceable.

What will serve us better than traffic in the future, the way traffic served us better than the horse a century ago?

Markets respond to values and costs. Make something easy and low-cost and you can be sure that most people will choose it. That is traffic in America today: the lowest-cost, highest-value option. Because there is no guarantee that traffic will remain the lowest-cost option, we must begin looking seriously at better solutions. Not in a revolutionary but an *evolutionary* way: a way with a clear path from here to the next thing.

The assumptions behind the traffic network and its infrastructure are starting to fray. The 2012 transportation funding law focused more on maintenance than new visions or construction. Per-capita driving fell to 1995 levels in the mid-2000s. Millennials and Gen Z are finding more opportunities in walkable cities than in drivable suburbs. The rise of COVID did not spur a return to traffic so much as a renewed migration to the internet, with online collaboration tools changing from niche to mainstream. America's transportation landscape and connected land use has changed in my lifetime and will keep changing. The only consistent pattern is that Americans are searching for something better than traffic dominance. Traffic may or may not be the dominant form of transportation in the coming decades, but it is time to start looking seriously at how America could evolve from traffic-only to transportation choice.

By 2010, per capita traffic use was down to 1995 levels and continuing to fall (see figure 1.1). Traffic congestion, often caused by the last few cars to get on the road, has fallen to similarly nostalgic levels (see figure 1.2). The largest segment of people not driving is the sixteen-to-thirty age group, the first, and riskiest set of drivers. They are finding ways to live without or with less driving. Future generations are becoming less attached to the notion of traffic as the sole means of transportation.

Figure 1.3 shows that the average number of miles traveled per year imperfectly follows the percentage of population employed.

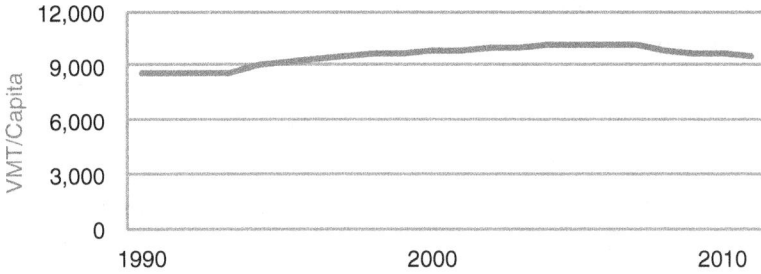

Figure 1.1. Per capita vehicle miles traveled (VMT), recently down to 1995 levels.[2]

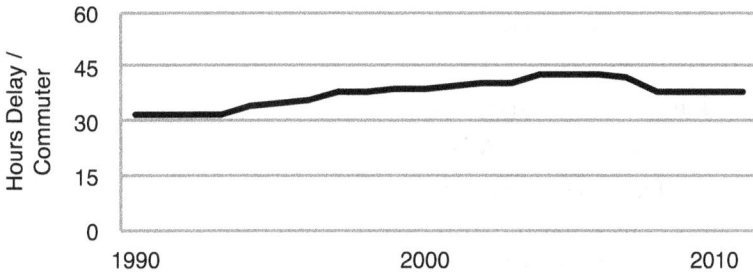

Figure 1.2. Annual hours of delay per commuter, down to 1997 levels.

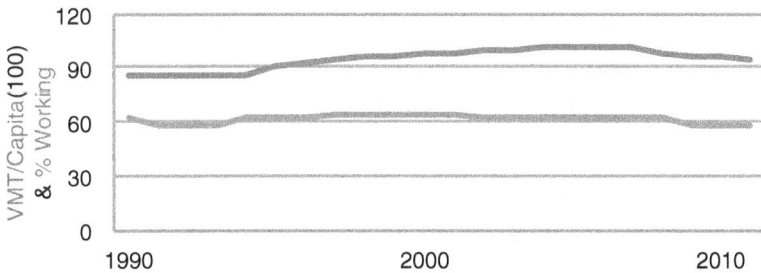

Figure 1.3. Comparison of vehicle miles traveled (VMT) and percentage of Americans working. Both decreased between 2008 and 2011.

Have we reached peak traffic? Since our new generation of workers and householders is not driving nearly as much as our forebears, won't we continue embracing human-scaled transportation as we continue to work and start families of our own? The recent downturn in traffic was caused more by a lack of jobs for this generation than a fundamental shift in our infrastructure or attitudes. A car or truck in traffic remains both signal of and method of getting wealth in much of America. We have already seen traffic volumes rise and fall with the economy, COVID, and gas prices in the first two decades of the 2000s.

Transportation and the economy are imperfectly linked. Every time the economy booms, people have more jobs and more places to go. Usually this means more traffic. With wealth, the demand for livable communities increases, and more developments and even roadway projects are built with accommodation, if not encouragement, for walkers and bikers. When the economy ebbs or when vehicles become expensive to own or operate, people have fewer jobs and travel less. Their trips often switch to local walking and biking trips.

Walking, biking, transit, and traffic all ebb and flow with the economy. But how do we develop an America where walking, biking, and transit are as much an engine of the economy as traffic?

**I GREW UP IN A STREETCAR SUBURB OF ATLANTA, WITH BUNGALOWS ON THEIR ONE-**
fifth of an acre, in blocks within walking distance of a paved-over streetcar track that hadn't carried anything for thirty years. When I was a kid, the bus came to that same corner bearing the same number as the trolley and the trolleybus had for sixty years before it. Growing up near Little Five Points and Decatur, as a teenager in love with my city, I saw what good walkable towns could be. Though I had no sense of its history, I knew there was once a walkable, bikeable transit city where my Atlanta was today. If I didn't grow up in a once-walkable part of Atlanta, I would have had no hope of biking it and getting around to everything I needed until my early adulthood.

Three miles out from my house, the conditions for bikers became downright dangerous. My teenager's journey out to Stone Mountain or Duluth was fraught with the hazard of unwitting drivers, driving cars that weighed ten times more and sped four times faster than me.

I recall seeing all kinds of grim wrecks on my commutes, some involving walkers. When I lived in midtown Atlanta, I saw the corpse of a woman splayed on US 29. Her groceries were the comet trail of her attempt to cross a busy highway an eighth of a mile from a crosswalk. As a kid, I choked down breakfast one morning while a dog died in the street, hit by some long-gone car.

As my Atlanta expanded with longer bike trips, I occasionally glimpsed streams under little bridges just as I started biking up another hill. In college, I came to closely study and get to know the fish diversity, fish size, and abundance of turtles, ducks, and other hidden life in those streams. I came to my profession, and ultimately to this book, when I saw how transportation choices affected the health of those streams—first through conservation biology and then as an urban planner.

I did not get my driver's license until I was twenty-two because I could bike everywhere in my native Atlanta. I sometimes had to bum rides from people to get to places quickly, but I was able to do many of my commutes and errands by bike. I didn't buy a Volvo 245 until I went to graduate school at Clemson University, 120 miles from home.

While at Clemson, I discovered that I was a city boy who loved Atlanta, and my Volvo was an essential means of escape back to that city life. I went back nearly every weekend that first year: four hours, no air conditioning, windows down in the summer, great Scandinavian heat in the winter, radio blaring. For a couple of years, I drove thirty miles every day to and from Clemson, just to live in a walkable part of Greenville: I was only able to live walkably with the aid of my car! During my first summer collecting field data, the Volvo served as transportation and motel. A 1985 Volvo wagon can hold a 4x8 sheet of plywood inside, or a twin futon with room to spare for luggage—better than most SUVs. I put 100,000 miles on that Volvo between 1995 and 2004 before its electrical system failed.

Atlanta is not known for its walkable urbanism. Growing up and moving around Atlanta and its metro showed me what traffic would do to a city for the sake of speed, distance, and parking. The historic trolley city I knew only as a ghost was replaced with expanses of parking and buildings that turned their backs on the street. I learned to walk and bike everywhere in deference to traffic. It is no coincidence that John Portman, the signature architect of the Atlanta skyline, made his name turning buildings inward to their beautiful atria and away from the street, leaving downtown Atlanta's streets devoid of life and only active at loading docks. Places like Snellville, Lenox Square, and Gwinnett County were endless washes of highway, parking, and subdivisions.

Traffic connected origins and destinations for millions of cars and trucks every day. While there was no point, place, or sense to walking or biking in these places, it was beguilingly easy to get anywhere you wanted in greater Atlanta as long as you were in traffic.

I have seen both Atlanta and America as a walker, a biker, and a driver. I also realized at Clemson that conservation biology and landscape ecology were ill-equipped to answer the question I had come to: "What do we do about traffic?" I asked this question in humility. I understood the easy freedom a car gave me that I would not have otherwise enjoyed. But I also knew the freedom and strength of being a biker amongst all the cars.

My studies at Clemson taught me an important lesson about multimodal transportation and land use. We were studying the movement of pollen carried by carpenter bees between patches of passionflower in an experimental landscape. It took me two field seasons of strange data to see that the large carpenter bees were flying through and beyond the experimental landscapes. They were too small. This is similar to the ways that a landscape built for traffic will be unusable for walkers, and a landscape built for walkers would be irrelevant to drivers. At best, a landscape built for walkers would be surrounded by a sea of parking for people from throughout the region to allow them to enjoy the rare treat of walking. You might recognize this as a shopping mall.

In the spring of 2020, the American model of commuting was upended by COVID-19. I drove to work that morning and came home midday. I didn't ride transit again for over six months. I never biked the long commute into work. I wasn't even sure we would ever commute the same way again. We already have been working on projects on the internet for decades. Video conferencing and chat bloomed in response to demand as office workers around the US found that physical presence at work was optional; even discouraged. As of this writing, I am working from home, for a job in an office that is over two hours' drive away.

As I got over the initial shock of moving from my office to my basement for work, I thought hard about the meaning of this book. What does it mean when the office and transit are disease vectors? What did it mean when the densest places in the densest cities looked like the places with the highest mortality? It seemed like the most socially distant way to get around was via traffic, together alone in our cars.

We knew nothing in that first year. We stocked up on gloves, masks, and disinfectant. Transit systems spent millions disinfecting every surface, including irradiating the handrails on escalators. We have since discovered that COVID is mostly spread by breath, not touch. We now know that COVID transmission is highest in the most crowded buildings, not the densest cities. If we avoid prolonged periods of loud, unmasked talking to people, we can greatly cut down COVID transmission. We can use transit, just be quiet about it. Soon I will no longer need my home office.

By April of 2020, we masked up and gloved up to go grocery shopping, and later to take walks in the neighborhood. We are still mindful to not pass too closely on the sidewalk. From May of 2020, two months after the beginning of quarantine in the US, I started to bike to the cafés that were open, get my coffee and breakfast sandwich, and write out on the patio as the world woke back up. I would get my writing done this way before biking back to my home office for a full day's work, connected but alone.

I resumed editing this book in earnest during a trip to New York City, one of the places that was worst hit by COVID. We had been through four pulses without getting it. It is becoming routine, but I still wear a mask into retail establishments and maintain a distance of at least four feet from other people in public. The standard was six feet when we were still in full unknowable panic about COVID.

This will pass. This is already passing. We learned a lot about what we will tolerate, and we will forget even more. The fact remains that America remains dependent on traffic, and could do with more choice in its transportation, land use, and housing policies. Can we learn to rebuild in a way better suited to the future, or will we wallow and founder in a century-old model of transportation and development?

THERE ARE SIGNS ALONG THE SIDE OF BUSY HIGHWAYS, PEEKING THROUGH THE WOODS over broad hills and crowded lanes, looking toward town, taunting "If You Lived Here, You'd Be Home Now." Every day, each of these signs challenges the diligent commutes, exurban lifestyles, and finances of thousands stuck in traffic, offering the simple succor of living closer to work. The sign offers the peace of less driving to thousands of bored commuters every day.

Would living there be as idyllic as the sign says? Living there would mean living next to the highway the rest of my days. I wouldn't want that—traffic was annoying enough an hour a day. Imagine going to sleep every night to the roar of a machine river, punctuated by the occasional screech of ruined lives. If I lived there, I might be home now, but if I drive just a bit further I can get to the lightly traveled capillaries of our hierarchical traffic system, far away from this multilane aorta. Though you can trade a long commute for a noisy apartment next to the highway, you can also gut it out and stay in traffic, both ways. About five hundred times a year, or about twenty thousand times in our working lives. Keep driving.

I was stuck in traffic one day, gazing, at 9 mph, at one of these "you'd be home now" signs, when it hit me: These signs would make sense even if they were much closer to town. Locations closer to workplaces could offer even less driving than the amount promised by this sign on this busy, remote, but still developed highway. What if, for every sign, there was another a mile closer, and then another? What if working and living were close enough that you could walk or bike there? How would that look, how would it work, and most importantly, *where* would it work?

**IN 2002, WHEN I WAS JUST GETTING INTO TRANSPORTATION PLANNING ISSUES, I SAW** a talk by Rick Hall, a member of both the Congress for the New Urbanism and the Institute of Transportation Engineers. After his talk on walkability, traffic dependence, and how to blend the two, he made a comment that has stuck with me to this day: "The guy with the simple job always wins." Building solely for traffic is much simpler in a landscape built out over the last century for the use of traffic. Road design, zoning, and parking standards have all been in place for that long. We have rebuilt America over the last century on the simple assumption that congestion is bad, and more lanes and more parking represent freedom. Most of walkable/bikeable America had already been built out before the rise of automotive traffic. Many of the most walkable and bikeable communities in America would be illegal to reproduce, and there is always pressure to bring them "up to code." Even though traffic engineers know about complete streets and multimodalism, it is simpler to get a road widening approved by most state departments of transportation than a multimodal road diet or walkable street grid. So that's what gets built. Planning and building for multimodalism are far from the default option. They are not nearly as simple as traffic dependence. We must identify a simple job to do, and how to do it well.

Streets, roads, arterials, and highways designed for traffic organize America's transportation and land use. Our homes, workplaces, and buildings are defined by a street address and have been for the last century or more. Not much has changed in the way we buy houses, shop, or get to jobs to make America a place that walks, bikes, buses, or travels by train—traffic is still the default. Adding transit services or routes is a years-long deliberative process, with plenty of opportunities for citizen engagement, objection, and veto. Bike lanes and sidewalks only recently (1992 ISTEA) became a federal funding priority, and most state departments of transportation (DOTs) see them as ancillaries. In contrast, adding traffic lanes to roads is routine, with more peripheral citizen involvement. We have not been very good at building new places without the assumption of traffic. Can we do better than this in the future?

In transportation planning, from road widening to the redevelopment of transit station areas to improve walking and biking access to the station, there is a jurisdictional tendency to only work within the right of way. These solutions tend to focus on fixing the area immediately next to the roadway to invite bikers and walkers, but they ignore the fact that bikers and walkers also require destinations they *want* to get to and are able to do so within a sensible proximity. Transportation planning remains separate from land use planning for many projects, even though they rely on each other and make assumptions about each other to make any sense of their scopes of work. Due to the distance between destinations in Atlanta, I would usually bike more than ten miles a day for commuting and errands. This is beyond what may be accessible or perhaps even advisable for an average trip, which I define for the purposes of this book as a three-mile radius for bikers or one-mile radius for walkers.

Different authors in transportation or land use support their arguments by choosing the local, municipal, or regional scope that supports the story they want to tell. In this book, I want to take a broad look at the transportation and land use of the US, without cherry-picking the cases that support my view. I will use the census block group scale, which is surveyed for the whole US, but is sufficiently local to be traversed on foot or bike. The data I choose is broad (the whole US), but detailed (block groups around all 4,096 rail transit stations). My purpose here is to make a national case for transportation diversity assessing the four primary modes of transportation in America—walking, biking, transit, and traffic—to arrive at a workable solution.

Traffic spreads everything out to an unworkable scale for walking, biking, and transit. Catering to the requirements of traffic affects how space is used and how people live in that space. The only places in America that are walkable and bikeable were either built long ago or are in single-lot subdivisions, surrounded by traffic. If you want to get anything larger than a coffee in these great urban enclaves, you need to get in your car. And woe betide the vendor of shoes or sweaters at the center of these walkable developments: they are obvious stores to go to for the people living in their subdivisions, but hard to reach for everyone else. If a walkable/bikeable "here" is to be practical, it must be surrounded by an awful lot of itself. Manhattan works because thirty-three square miles of it were built between 1620 and 1900 with walking in mind. Philadelphia, with its townhouses and corner stores, works because of a similarly long period of walkable urbanism. Even my native Atlanta has walkable cores that were towns once served by trolleys and are now surrounded by highways. Where can we build *new* walkable places, if not as isolated subdivisions in the countryside?

The key to letting all transportation compete with traffic is land use and proximity. Traffic exacts a toll on property owners in the form of parking: if you want to build near a road and connect to that road, you need to use your property for parking. What we don't require is that if you want to build near a transit station, you should be able to build for walkability and bikeability to connect to that transit station. Bring doors closer to each other and to transit, and let biking, walking, and transit compete with traffic and offer real choices in American transportation and land use. Right now, this is possible, but it goes through a long planning process of variances, public meetings, and citizen objections. It is much easier to build and expand arterials and put in new, single-use subdivisions designed for traffic. There are plenty of books that lay out complete visions of how we could do better, but not many that ask what single step we can take to make transportation choice real in America. This book is just a start, offering one simple prescription: let transit, biking, and walking affect land use as much as traffic does, and let them compete with traffic. If traffic requires that land use be dominated by parking, what if transit required that land use be dominated by places within walking and biking distance? If this new alternative does well and offers more prosperous and popular development for less money to people and governments, then other places too are free to build their own transit networks to awaken their land use potential. Again, the idea I lay out here affects only 1 percent of the United States, which is a smaller area than is paved by roads and parking today. The rest of America can develop as it has, in traffic. If transportation choice is appealing and is more prosperous in the twenty-first century than traffic was in the last, then communities are free to build with transit, walking, and biking in mind.

> Let transit, walkability, and bikeability affect land use as much as traffic does, so they can be as usable for transportation as traffic is, using only 1 percent of the area of the United States.

This is not a book of case studies or design solutions, but a geographical and statistical argument in favor of transportation choice in America. If the scale of walking and biking is not the same as traffic, where can we rebuild to enable this scale?

This is a book about the unrealized benefits of transportation choice to America. This book identifies where in America is already walkable, and where it makes the most sense to be walkable. How can we build places that work for walkers and bikers as well as transit users?

What if we could build places where we didn't have to use our cars in traffic every day for every trip? What if we built places where the obvious way of getting around was walking or biking from our front doors to work, play, and shopping? What if we only needed our cars twice a week, rather than twice a day? Proximity to where we want to be restores our time and choice in transportation.

Where *does* it make the most sense to build for walking and biking? We already know *when* it made the most sense to build for walking: before 1920. Places built then were built to be walked through, seen, and used by people moving at 3 mph. Now we build places to be driven through, glanced at, and used by people moving at 31 mph. Walking, biking, and transit are all relegated, restrained, and omitted by the traffic-dominated landscape.

This is not a new call. The Smart Growth and New Urbanist movements are just recent incarnations of skepticism about traffic that stretches back to the beginnings of the motor vehicle in traffic. Smart Growth and New Urbanism are both movements from the last quarter of the twentieth century that challenge the dominance of traffic in transportation and land use policies. Smart Growth's approach is more regional and functional while New Urbanism is more local and design-oriented. Both innovate on not just transportation and land use, but also housing, parking, watershed management, resiliency, and economic equity. The modern arguments against traffic have been functional or aesthetic. These protests have been drowned out by the obvious convenience of traffic. Nothing else provides safe door-to-door trips to most of America. To prevail, Smart Growth and New Urbanism must provide a *better* alternative to traffic, not just a prettier one.

The answer is in proximity, scale, and ease. The reason walkable places succeed is that they offer enough work, shopping, and things to do within walking distance of where people live. Simple, but nothing like the way we have built for the last eighty years. To make a place walkable, you need a lot of walkability around it. People need a reason not to just sit in their car for an easy ten-minute drive to the café or drugstore. Conversely, businesses need enough workers and customers within walking distance to make a place walkable, otherwise walkable places are little more than shopping malls with the parking lots behind the buildings. You need a critical mass of walkable area for it to work, and you need it to be more prosperous for less money than strip malls behind parking lots in front of residential subdivisions if it is to outcompete traffic. Walkable, bikeable places defy development standards written with only traffic in mind. For walkability and bikeability to work, we need development standards that enable walking and biking as routinely as road or parking standards enable traffic. Where should walkability and bikeability be the default, rather than the variance? In this book, I will make the case that America already has the tools it needs to reintroduce transportation and land use choice into its most prosperous metros by enabling walking and biking within biking distance of our installed base of 4,096 rail transit stations.

This book is about traffic, as well as getting over traffic. I see traffic as a great and wonderful tool. I also see it as a liability for America's future if it remains our sole mode of getting around. This book is an appreciation of traffic as the descendant of thousands of years of transportation evolution and serves as a proposal for the next step along that evolution.

The next chapter looks at the history over the centuries of America's national and local transportation and land use, so we can understand how we arrived at our current situation.

# 2

# How We Got Here

To propose next steps in America's transportation, we should know how far we have come. This chapter is written in several threads. Humans and their ancestors have been walking on two legs for over four million years.[3] Other modes are more recent, with transit (350 years),[4] traffic (250 years),[5] and biking (200 years)[6] the sole province of the wealthy for decades before they were commercially affordable. Before 1800, the area of the United States was fragmented relative to more prosperous and fertile South America. The emergence of America as an economic leader—and a unique advocate of traffic—is instructive, and may hold the keys to our success in this century.

## Walking—from prehistory to 1800

America was populated by Siberians walking and expanding across the Bering Strait land bridge between twenty thousand and ten thousand years ago. South Americans—such as the Maya, Olmecs, and Aztecs—traded crops with the people of the United States over two thousand years ago. Most of our major crops and livestock were first domesticated in South or Central America.[7] The United States was never unified under an empire, like the Inca or the Aztec empire, but was populated by a diverse set of distinct nations between the coasts, woodlands, plains, desert, cold, and mountains, each with its own customs and policies toward its neighbors.

Over millennia, four hundred nations within the area of the United States cleared and maintained a network of trails—like the Onaneechi Trail (modern I-85),[8] the Iroquois Road (NY 5),[9] the Pequot Path (US 1),[10] and the Sauk Trail (US 12)[11]—using bark-stripping, fire,[12] and foot traffic to create trails no wider than today's Appalachian Trail. Extrapolating from the thirty thousand miles of known trails in the east (figure 2.1), there were possibly as many as one hundred thousand miles of trails connecting nations to each other, a distance comparable to our US highway system.

Europeans came to America in search of trade, not plunder, but plundering came naturally to them. In the century after the Black Death,[14] Europe was a growing, warring, debt-ridden place,[15] with Spain looking for cash from wars within and without its borders. The peasants and city dwellers of Europe would be subject to war, draft, sacks, and pillages until 1945. In that regard, the way the Europeans treated the Americans they met during colonization was par for the course.

As many as sixteen million Americans occupied the area of the United States in 1500, but Eurasian diseases like measles, typhoid, and smallpox may have killed 50–90 percent of Americans before they ever encountered a European.[16] Americans had not dealt with these diseases for hundreds of generations and were tragically susceptible to them. In return, Americans gave Europeans syphilis, with 50 percent mortality rates for infected Europeans in the 1500s.[17]

In Central and South America, Aztec and Incan empires were usurped by Spanish conquistadors in the 1520s and 1530s. Spain imposed its language and hierarchy on millions of Americans at once, creating "Hispanic" culture over the centuries. The history of the United States between 1500 and 1950 was the forceful, piecemeal, or accidental takeover of land from Americans of four hundred nations, erasing or displacing most of their cultures in the process. European colonists from three nations (Spain, France, and England), became Americans of one nation in the process (see figure 2.2).

The first colonist toeholds were isolated and extractive. The native American population fell constantly over the next five centuries and the Europeans rapidly outnumbered them (figure 2.2). The American suburban ideal was born in these first colonies. For generations, European towns in America were walled port cities. Only the crazy, clever, and brave ventured out to trade with the Americans or trap furs and meat for long, lonely weeks in hostile territory. The rest made their living trans-loading between the extractors and shippers. America's resources were worthless unless the colonists could

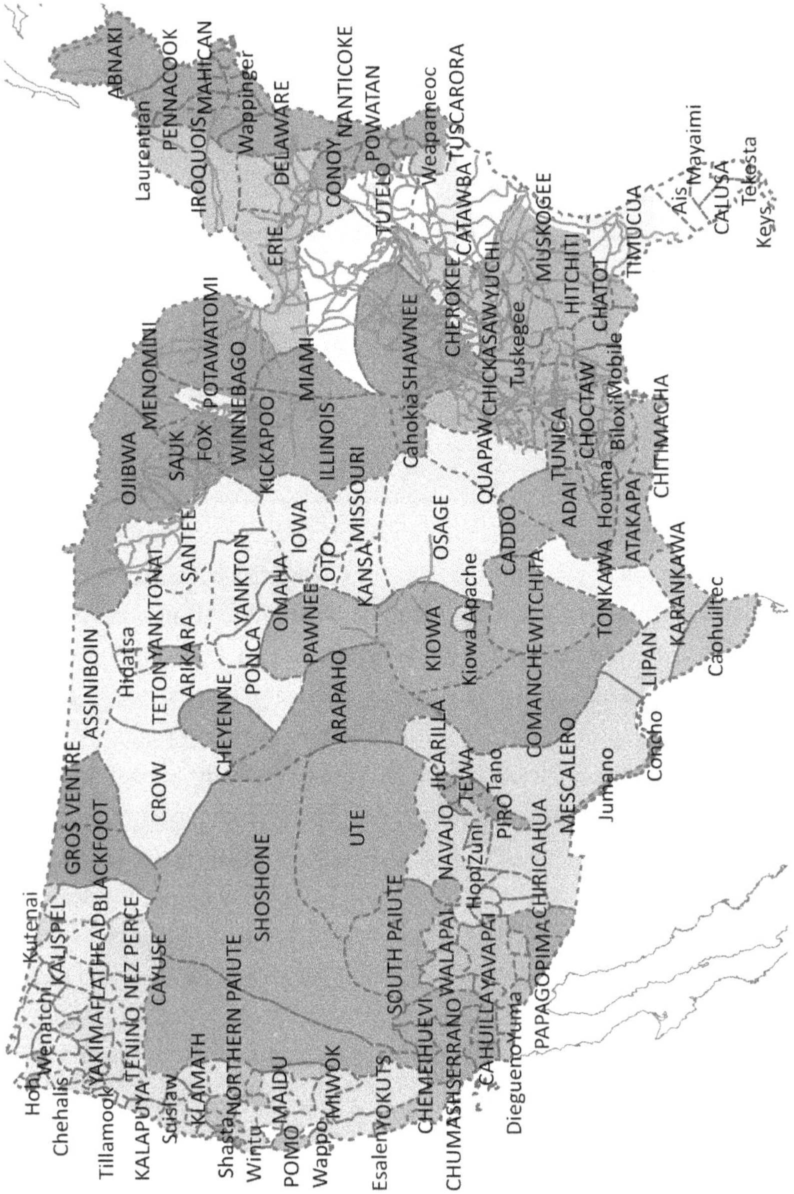

Figure 2.1. American nations, language groups, and known trails, ca. 1500.[13] Dotted lines are approximate boundaries between nations. This figure appears in color online.

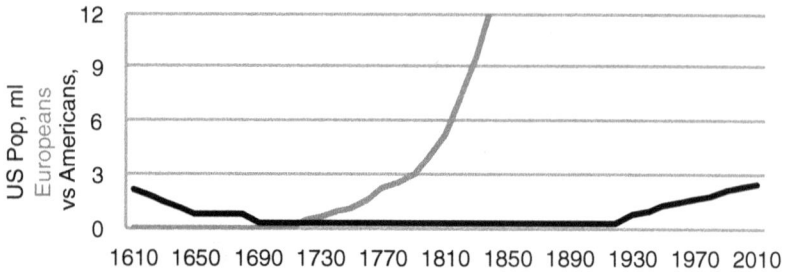

Figure 2.2. **American and European populations in the area of the United States, 1610–2010.**[18] "ml" = "millions." Black line shows the American population, while gray line shows the European population.

sell them back to Europe for money. Commerce between the colonies and cities was at first negligible and only took place by ship.

It is hard to know the walking history of America because it was merely the everyday activity that people did between other things. People lived their recorded lives in places they walked between. Our coastal and river cities were walking cities until the early 1800s, and homes, docks, gardens, pubs, tanneries, and slaughterhouses all had to be within walking distance of each other. The horses that were in these cities were for hauling freight, wealthy passengers, or heavy lifting. People walked freely everywhere, except for women. "Street-walking" (literally a woman just walking unaccompanied on the street!) was considered advertisement of prostitution in American cities until the twentieth century. This arrangement was simple for customers, police, and predators,[19] but dangerous for any woman who had to walk outside.

Colonial Americans usually walked to get around, not for the sake of exercise. Streets were disgusting. There was no system of trash or sewage management in any American city before the 1860s. The streets were open sewers, with tiled trenches in the center of nicer streets to carry the morning's sewage away to the river. The first underground sewer network wasn't put in under Boston's streets until 1876,[20] and the first sewage treatment plant wasn't built until 1890 in Worcester, MA.[21] Drinking water was always collected upstream of the town's outfall. This worked as long the city didn't grow to begin drawing water downstream of its outfall.[22]

The lack of sewage management resulted in disease and death rates that are shocking by today's standards. Diseases almost unheard-of in the western world today, like tuberculosis, malaria, and cholera, killed thousands each year, as recently as 1900. Table 2.1 shows the change from contagious diseases to chronic diseases and the dramatic decline in death rates since 1850.

Table 2.1. Top ten causes of death in the US from 1850–2000, showing the change from catastrophic to chronic diseases.[23]

| Rank | 1850 | 1900 | 1950 | 2000 |
|---|---|---|---|---|
| 1 | Tuberculosis | Pneumonia | Heart Disease | Heart Disease |
| 2 | Diarrhea | Tuberculosis | Cancer | Cancer |
| 3 | Cholera | Diarrhea | Stroke | Stroke |
| 4 | Malaria | Heart Disease | Accidents | Lung Disease |
| 5 | Fever | Stroke | Old Age | Accidents |
| 6 | Pneumonia | Cirrhosis | Pneumonia | Diabetes |
| 7 | Diphtheria | Accidents | Tuberculosis | Pneumonia |
| 8 | Scarlet Fever | Cancer | Arteriosclerosis | Alzheimer's |
| 9 | Meningitis | Old Age | Kidney Fail. | Old Age |
| 10 | Pertussis | Diphtheria | Diabetes | Poisoning |
| Deaths/1,000 | 20 | 17.2 | 9.6 | 8.7 |

Professional sweepers swept sewage along the streets for tips from walkers desperate to get to the other side.[24] Pigs roamed the streets, feeding on the ubiquitous piles of food scraps, and manure (horse and human). Until the 1880s[25] there was a vigorous market in horse manure to local farms, which closed the loop nicely, considering that at their peak horses ate hay from an area of farmland the size of West Virginia.[26] Horse urine (a quart daily per horse)[27] was perfect feedstock for tanneries, among colonial America's first industries.[28] The trade in horse manure continued until the late 1800s, when supply overwhelmed the ability to cart it away.

The first municipal trash collection in the US was in 1866, in New York City.[29] The first modern sanitary municipal landfill was built outside of Fresno in 1937, the brainchild of Jean Vincenz.[30] Before then, cities with a waste collection program took every day's trash to a dump, with ubiquitous ragpickers and crows completing the work of the pigs back in town. In cities without a waste program, every building had its midden for the pigs.[31]

You can imagine the smell. The poorest neighborhoods with the cheapest rents were downwind of all this, and the very cheapest rents were right next to the worst industries: factories, mills, and stockyards. Being close to the industries shortened worker commutes, and lives. The factory owners and wealthier traders could afford to live upwind and away from the grimy seats of commerce. (Decades later, this "favored quarter" still gets much of the infrastructure and institutional spending, while the other three quarters struggle for attention and even for postindustrial jobs.[32]) Most workers and residents in these cities of sewage, trash, and smoke dreamed of getting away from the city as often and as much as possible. Until high-speed, low-cost transportation became available, escape was a rare treat.

Many of America's founding fathers could not stand the city. George Washington was happy surveying large tracts of countryside for farmsteads or new port cities like Alexandria. Thomas Jefferson worked on the Declaration of Independence as far out in the suburbs as he could find lodging, at Seventh and Market Streets.[33] As governor of Virginia, he set up the state's city-county distinction to insulate agricultural counties from corrupt and overcrowded cities.[34] His land ordinance of 1785 divided Indiana and Ohio into nothing but farmable lots. Jefferson's ideal citizen for the new democracy was not a city worker, but a farmer.

## The rise of transit—1800 to 1910

The roads that did exist before 1776 (figure 2.3) were expanded from American trails or hewn from the forest. These trails were little more than footpaths themselves, susceptible to the seasons and to rutting. Ohio did not explicitly outlaw leaving stumps in newly cleared roads until 1803.[37] The "Flying Machine," an unbelievably fast 1766 stagecoach between New York and Philadelphia, could make the trip southbound in three days (1.4 mph average speed)[38]—a journey you can now make in less than three hours by traffic or railroad.

Figure 2.3. The first map showing roads in America, 1690, by Christopher Browne.[35] Provided courtesy of the Norman B. Leventhal Map & Education Center at the Boston Public Library under a Creative Commons Attribution Non-Commercial Share Alike license.[36] This figure appears in color online.

Stagecoaches, switching teams of horses every two to twenty miles on an otherwise nonstop route between two points, had been in use between American cities since the 1750s.[39] The first public transit[40] routes in America were stagecoach routes within cities that were sprawled out by circumstance, such as Boston–Cambridge in 1793[41] and Georgetown–Washington, DC, in 1804.[42]

New York in 1820 was crowded. One hundred thirty thousand people lived ninety to an acre[43] in the 2.3 square miles south of Houston Street (SoHo).[44] By 1840, New York's population had more than doubled to 312,000,[45] but had expanded only a quarter in area north of Washington Square Park (est. 1827).[46] It was under these stifling conditions that America's urban walkers welcomed transit. Omnibuses, introduced to America in 1827,[47] differed from stagecoaches in that they required no prior booking and would board and alight passengers at multiple points along the route.

The omnibuses of 1828 drove in the same dirty streets as draft horses and walkers, streets that would turn to impassible mud with every rainstorm.

An 1850 traffic survey at Broadway and Chambers Street counted over sixteen thousand omnibuses and nine thousand other vehicles in a single day.[48] Horse-drawn omnibuses continued to improve throughout the 1800s, including point-to-point hackneys as taxis and passenger-focused "herdics" (small two-wheeled omnibuses with sprung suspension)[49] that made the services even more convenient. Per ride, point-to-point services could charge a 500 percent premium over fixed-route omnibuses,[50] a precursor to the price difference between transit and traffic.

Taking a cue from mines and new freight railways, transit companies in New York (1832) and New Orleans (1835) laid down rails along a couple of the most lucrative omnibus routes and changed their iron-banded wheels to flanged.[51] Smooth iron rails allowed much less rolling resistance than the dirt roads. While omnibuses could reasonably carry nine passengers, horsecars on rails could carry over a dozen.[52] The horsecar on rails was invented in America,[53] just four years after the omnibus was introduced.

Between 1832 and 1870, horsecars evolved from wagon-like omnibuses to streetcars,[54] at the same time as railroad cars were evolving from wagons to boxcars. Railroad trains, many times larger than horsecars, required special caution, care, and signaling.[55] The railroads asserted property "right of way" over their tracks, and guarded it against trespass. This was a precursor to the right of way adopted for most roadways in the twentieth century.

Rails in city streets caused new problems. They were the same rails with ties that railroads have today. Anybody walking, trotting, or rolling across the tracks could trip or get stuck if not careful. In New York, rails had to be plowed immediately after snowstorms, interrupting the flow of sleds and regular horse-drawn traffic that had served for over a hundred winters. As a result, horsecars on rails did not develop in America for twenty years, until Emile Loubat, looking to bring horsecars to Paris, developed a way of making recessed rails that could be installed flush with the street (figure 2.4),[56] and within a decade omnibus operators in Chicago, Boston, Baltimore, Pittsburgh, and Philadelphia all installed rails for new horsecar lines.[57]

Horses were the cheapest, most reliable, and most powerful source of energy and pulling force for centuries in American cities. Horsecar operators were in the transit business, not the stabling business. They ran as few horses as possible to cover their routes, year-round. Their horses lasted an average

**Figure 2.4. Horsecar and carriages in Washington, DC (1893). Rails are flush with the dirt road surface.**[58] **Courtesy of the Library of Congress.**

of two years in service before dying in the street or in the stables[59] (compared to a typical life of over twenty-five years for a horse today). The plight of these and other workhorses in New York was the motivating injustice behind the founding of the ASPCA.[60]

Horses were not a safe mode of transportation. In 1900, and for decades before, over one thousand people were killed on, near, and around wagons drawn by horses for every one hundred million horse-miles traveled.[61] In comparison, 1.13 people were killed in 2012 for every 100 million car-miles traveled. Most of the danger was to walkers kicked or bitten by the horses, not to the passengers, riders, or drivers.

Throughout their working lives, horses expended a great deal of energy. Each horse needed twenty pounds of hay a day to keep up with the schedule of horsecar or draft duty. At their peak population in 1915, an area the size of West Virginia was cultivated for hay for horse feed.[62] As a consequence of their appetites, each working horse excreted fifteen to twenty pounds of manure and one quart of urine, every day, onto city streets.[63] The sidewalk was used for centuries to separate walkers from sewage, not from traffic. After every rain the street would become an open sewer. On dry days, the manure would dry out and blow around the city as dust, coating the town with the stench and disease of horse waste.

The equine influenza epizootic of 1872[64] stressed America's dependence on horses.[65] America depended completely on horses to move the economy. Horse-drawn wagons delivered the coal fuel to power steam locomotives, and horse-drawn wagons completed all deliveries to and from railroad stations and ports.[66] During Boston's 1872 Great Fire, firefighters had to drag firefighting equipment and water tanks between each fire by hand, because no horses were available due to influenza.[67] The great epizootic contributed to a worldwide depression that lasted until 1879.[68] It was a signal that our economy needed something better than the horse.

In the 1870s, no motor vehicle was more effective and affordable than the horse. Streetcar inventors tried naphtha gas (Brooklyn, NY),[69] natural gas (Canton, IL; Cicero, IL; Elizabeth, NJ; and Paterson, NJ),[70] batteries (Sacramento, CA;[71] New Orleans, LA),[72] ammonia gas (New Orleans),[73] and even compressed air (Rome, NY, in 1900),[74] but none were as cheap and reliable as the horse. Miniature steam locomotives pulling trailers were the most successful alternative, but the noise and steam annoyed neighbors and scared horses. The streetcar companies, thinking it was the appearance of locomotives scaring the horses, decorated many of them with paneling to look like streetcars.[75] This did nothing about the noise and smoke from the steam engines, however. The horses and the households along the routes just had to get used to the noise and exhaust of steam engines passing by every half hour.[76]

After witnessing a team of horses pulled backwards to their deaths down steep Jackson Street in San Francisco by their overloaded horsecar,[77] Andrew Hallidie adapted mining cable-car technology for use on the Clay Street line.[78] His first successful attempt was the cable car, towed by cables running in a channel embedded in the road. Getting the power source away from the streetcar reduced vehicle weight and power needs, while allowing the transit company to manage just one large, efficient, and central power plant in a controlled environment. The cable car was also successful in flatter cities like New York and Chicago. It was adopted in twenty-three other cities between 1882 and 1890.[79]

Cable-pulled cable cars were best suited for straight runs up hills. When the cable car had to turn through curves, the operator had to release the cable grip from the first cable and let the cable car coast to the next straight run of cable, which they could grip back onto and then proceed. Braking or hitting anyone in this curve would have meant they had to get the car pushed to the next cable run, so the operator would frantically have to ring their bell to warn everyone in the street.[80] If cables became frayed with use, the grips could be caught in them, dragging the cable car to the end of the line. These were not high-speed accidents—they took place at less than 10 mph—but they were still dangerous.

Streetcar inventors had long seen the potential of electricity but had not managed to make it work. By the late 1800s, electric motors were doing useful work, but they were still not suitable for streetcars. The problem was getting sufficient power to streetcars without electrocuting passersby. Inventors like Daft, Van DePoele, and Bentley-Knight all had their preferred methods of power delivery. Leo Daft started with a lower-voltage system along a publicly exposed third rail but moved to dual overhead wires with an overhead "trolley" to draw power from the wires. Charles Van DePoele used an overhead wire from the start, using a spring-loaded pole to make contact between the streetcar and its power source. Bentley-Knight worked mostly with conduits below the ground, taking a hint from cable cars to contact a charged wire with a conducting shoe to supply the streetcar motor.[81] These pioneers worked out the steps toward electric trolleys in Baltimore, MD (Daft, 1885),[82] South Bend, IN (Van DePoele, 1885),[83] Montgomery, AL (Van DePoele, 1886),[84] and Pittsburgh, PA (Bentley-Knight, 1887),[85] but these were all single lines with few cars and frequent outages. In 1888, Frank Sprague first demonstrated a network in Richmond that could reliably supply power to fourteen streetcars at once. His system used the rails for return current (disturbing users of the equally new telephone).[86] Sprague's system for Richmond was the first system to offer full-scale service in a large, thirteen-mile trolley system, after his smaller attempts in New York and Missouri.[87] In the year after the Richmond trolley demonstration, electric trolley service grew from forty-eight miles in thirteen cities to over one thousand miles in 110 cities.[88]

By 1900, trolleys offered the only way of commuting long distances for many Americans. The first rail suburbs, like Llewellyn Park, NJ (1853)[89] and Riverside, IL (1875),[90] were built along rail freight lines to serve families wealthy enough to afford daily passenger train tickets to and from the city.

The appeal was refuge from the noises and smells of downtown, the sole privilege of the wealthy up until then. "The good life" required getting away from the crowds and hazards of the city.

With newer, cheaper electric trolleys, new towns and subdivisions sprawled on the edge of town around trolley stops to exploit that demand for suburban living. Neighborhoods like Inman Park, GA (1896),[91] Chevy Chase, MD (1890),[92] Cleveland Heights, OH (1892),[93] Maplewood, MO (1900),[94] Roland Park, MD (1901),[95] and Coral Gables, FL (1925)[96] sprang up between 1886 and 1929 to appeal to upper-middle-class workers who could finally afford housing away from the stench of the city.

With the development of the trolley and the affordable delivery of power to streetcar motors, transit providers expanded beyond single-city systems. Starting with two-city systems like the Twin City Rapid in Minneapolis-St. Paul (1890)[97] and city-suburb networks like the East Side Railway in Portland (1893),[98] interurbans were built through the countryside between cities, with hundreds of routes covering dozens of miles each by 1910. The reason Los Angeles is spread out, yet so dense is the long-gone Pacific Electric "Red Car" network, which served a vast area between San Bernadino, Reseda, Venice, and Balboa. A precursor to today's light rail,[99] Los Angeles transit competed for range and speed with the steam railways, including freight service. Many of the routes laid down by the interurbans in the first decade of the 1900s are still in use as state highways or short line railroads.[100]

Like turnpikes a century before, interurbans had a hard time making money, no matter how miraculous they were for farmland and suburban transportation. Interurbans were converted one by one to regional bus carriers between 1920 and 1950. After deregulation in 1982, most of the unprofitable routes were shut down altogether.[101]

To separate mass transit from the crowd of foot, horse, and wagon traffic, America's most crowded cities built elevated railways. Starting in New York (1867),[102] Louisville (1886),[103] and Chicago (1892),[104] several cities developed elevated railways to let transit move independent of the crush of traffic below. Unfortunately, elevated railways turned once busy and popular thoroughfares into caves of ashes, cinders, and sparks. At first, elevated transit trains were pulled by a single locomotive. Later, another one of Frank Sprague's inventions, Multiple Unit Control, provided coordinated power for each vehicle in the train.

Cities with enough wealth and congestion built their highest-capacity transit lines underground, such as Boston in 1897, New York in 1904, and Philadelphia in 1907.[105] The concept of underground rail lines was already proven with Brooklyn's Atlantic Avenue Tunnel in 1845,[106] London's Underground in 1863,[107] and Manhattan's Pneumatic Subway in 1870.[108]

## THE DECLINE OF TRANSIT SYSTEMS

As transit evolved between 1804 and 1900, it required more capital, more centralization, and more coordination with city government, though transit was not usually a government-run agency. Omnibus, horsecar, and even some early cable car lines could be run as independent businesses, but as the fixed costs (and the marginal returns)[109] of transit rose, they consolidated into local transit monopolies. These city-wide transit monopolies were not uniformly beloved by citizens, passengers, or politicians, and were held responsible for service or street imperfections.

Local monopolies controlled the timing and service of transportation for entire cities and were well-regarded only when they provided excellent service. The mood was good when transit offered the fastest, easiest service in crowded city streets and the only access to faraway places and new resorts on the edge of town.[110] This required continuous improvement and maintenance; otherwise they were as much of a threat to the public as railroad "robber barons," like Gould,[111] Vanderbilt,[112] or Harriman.[113] Trolley barons like Yerkes (Chicago),[114] Huntington (Los Angeles),[115] Whitney (Boston),[116] and Widener (Philadelphia)[117] were more powerful, and more reviled in their cities than the nationally famous robber barons of the day. They were even more hated in the dozens of cities where they had an absentee interest in the transit systems. They built as shabbily as possible and ran their routes as crowded as possible. Then, as today, they made more money on fares when occupancy was highest. Packed streetcars were a boon for the operators, whose byline was "It's the straphangers who pay the dividends."[118] The excesses of these monopolies in the 1880s and 1890s made them a public target. Undermining and dissolving their trusts was considered a public good, resulting in progressive legislation[119] for alternatives to transit.

The typical fare for a horsecar, cable car, or trolley ride was a nickel well into the 1910s. Between 1865 and 1895, America's economy was deflationary, meaning that the value of a nickel was rising every year. Transit operators were getting more money every year with their fares, even if ridership didn't change. This was such a great deal that many transit companies agreed to nickel fares "in perpetuity," which looked like a great deal until the value of a nickel peaked 1899 and fell afterwards with inflation (figure 2.5). With collapsing budgets, transit companies had to cut corners—on expansion, service, hours, staff, and maintenance. By 1925, the tracks and cars of many transit companies were showing the wear of decades of deferred maintenance.[120] The omnibus, now motorized, was beginning to make a comeback on newly paved roads. By 1949, bus transit trips outnumbered rail transit trips.[121] By 2018, rail transit and bus transit had rebalanced, with approximately the same number of trips on each mode.[122] COVID fundamentally changed the transit market starting in the spring of 2020, and it remains to be seen how transit will change in response to the pandemic.

Another challenge to private operation was the antitrust case against electric utilities that ran streetcar companies, such as Georgia Power's Atlanta streetcar network. While many power companies were founded to supply power for trolleys, they had since become much more profitable, supplying power to homes and businesses. They kept running the streetcars as a loss leader and signal of good community relations. By 1935, the federal government forced electric utilities to focus only on power generation and transmission, leaving transit to the transit agencies.[124] This forced many trolley companies to switch to lower-cost buses, to ask the government for aid, or to go out of business.

By the 1960s, many transit companies had switched to cheaper buses and were still facing ruin. The road network, built to provide democratic traffic against monopoly-held transit, had succeeded in its mission, much as the Interstate Commerce Commission broke the back of the private passenger and freight railroads.[125] It was clear that not all Americans could afford a car, and America's roads could not accommodate all Americans driving on them. The first federal aid to transit was in the 1961 Housing Act.[126] By 1970, almost all transit companies had taken the bait, and had become public agencies. They

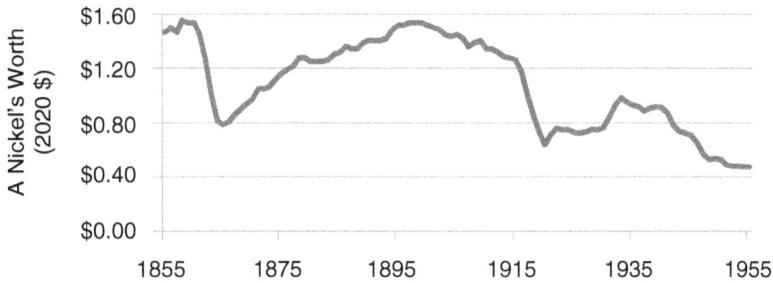

Figure 2.5. The value of a nickel as measured in 2020 dollars, 1855–1955. Note the rapid periods of inflation during the Civil War (1861–65) and World War I (1914–18). Deflation in the late 1800s occurs as the economy outgrows its money supply.[123]

were no longer explicitly in transit for the money, and they were hungry for ways to make transit the most appealing way to move around their cities. Although transit systems persist in these cities, most have been converted to buses that operate on streets now designed for cars, resulting in diminished transit ridership and indirect bus routing (figure 2.6).

## The growth of biking

Urban congestion for the first half of the 1800s inspired both transit and the use of bicycles. It took bikes longer to develop through the 1800s, with many uncomfortable and esoteric intermediaries, but transit and bike shared a golden era between 1880 and 1900. The first prototypes of bikes were built and sold around the same time as the first experimental cars, around the turn of the nineteenth century. Scooters like the "Dandy Horse" (1817) were novelties for the wealthy to glide faster than walkers or horses.[128] These fifty-pound wood prototypes of the bicycle featured two equal-sized wheels, a saddle, and a steering handlebar with similar dimensions to today's bicycle, but it was basically a two-wheeled wagon. To move, the driver would kick their way along the road to roll down hills and over flat terrain. Treadles attached to the rear axle were added for propulsion by 1845 (figure 2.7).[129] The bike became much more popular after the addition of pedal-driven front axles, with the "Boneshaker" of 1863. The Boneshaker's wheels still had straight iron or

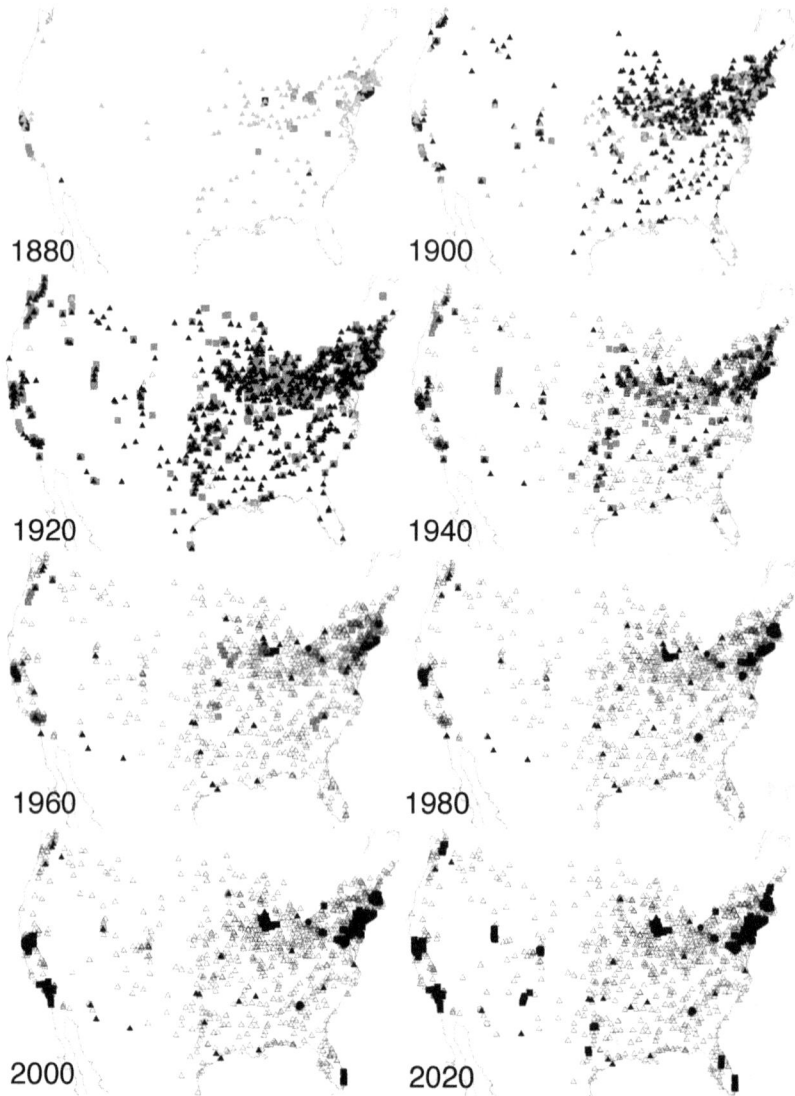

Figure 2.6. Transit systems 1880–2020. Horsecars are the primary mode of transit until 1890 when trolleys begin to take over. Trolleys dominate but begin to be supplemented by subways by 1910. Buses take over as the primary transit mode in 1940.[127]

wood spokes with solid steel rims. These wheels conveyed the road's texture directly to riders' sore bodies. Despite the discomfort, the Boneshaker was still the most convenient and popular form of the bicycle yet. Mania for the developing bicycle spread among those who could afford them in Paris, London, and New York in 1868 and 1869.[130]

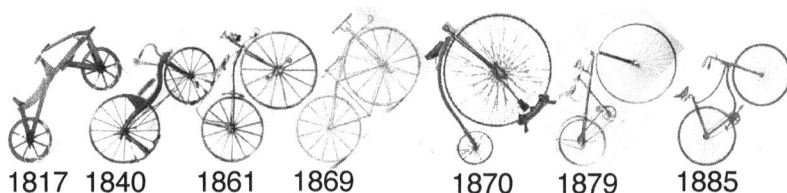

**Figure 2.7. Nineteenth-century development of the bike in France and the UK.**[131]

To get more speed out of pedals attached to the front wheel, English builders like Starley[132] enlarged the front drive wheel[133] and incorporated shock-absorbing tangential wire spokes[134] with his Ariel or "Ordinary" model in 1870. Albert Augustus Pope copied this English model in America with his "Columbia" bike in 1878.[135]

Bikes were faster than anything else their size when they were most popular in the 1890s. They offered unprecedented speed and freedom of movement. They were also the first self-driven vehicles on the streets—everything else was either pulled or on rails. The only reliably hard places on most city streets were the sidewalks, so cyclists traveled on these, which did not please walkers. The first streets paved for bikers were within city limits, but usually along riverbanks on gateways into the countryside. These were roads like Montgomery Avenue on the Schuylkill in Philadelphia, the Coney Island Cycle Path[136] in New York, and the California Cycleway[137] in Los Angeles. Bikers saw this as an opportunity to change American roads in the image of bikes, and started the "Good Roads Movement" and the "League of American Wheelmen" in 1880.[138] These were backed by the bike manufacturers themselves to enable the easy use of their products.[139] The first national highway department, the Office of Road Inquiry, was formed in 1893 under the Department of Agriculture, with the express purpose of studying road paving for the sake of bikers and farm wagons.[140] Automobiles were not yet a consideration in 1893.

The innovation in bicycles balanced comfort and speed with sprung, tangential spokes, then large drive wheels, and finally the pneumatic tire. The dentist John Dunlop developed the first marketable pneumatic tire in 1887 from a loop of garden hose for his boy's bicycle after reading about a previous attempt by Robert Thomson.[141] The pneumatic tire spread the area of contact with the road depending on the weight of the vehicle and the pressure in the tire. Air trapped in the tire served as a spring for both rider and road. The increased rolling resistance of pneumatic versus solid tires was more than paid back by the reduced breakage of wheels, spokes, and people's bones on bikes and wagons. The pneumatic tire and hard pavements attended both the mania in biking and later automotive markets between 1890 and 1910. This revolution in wheels and paving was abetted by the inventions of synthetic rubber in 1879[142] and synthetic asphalt in 1900.[143]

In 1850, most streets were just the spaces between buildings, and any paving in gravel, granite setts, or cobblestone was the responsibility of the adjacent property owners.[144] Many roads were just the dirt paths between buildings, wide enough to admit the occasional delivery wagon on a dry day. Streetcars were the first to interrupt this arrangement, but only on a few of the very busiest streets in the most successful towns. Horses and mules, the only source of motive power until the late nineteenth century, needed rough road surfaces for grip if they were going to pull wagons, omnibuses, or horsecars.

Paving had to change when the source of motive power changed. A car or bike pushes down with its own weight on its wheels, which are driving, not pulling, the vehicle. Instead of a rough road, a smooth, paved road is best for these self-driving conditions. Rails are an extreme case. Steel rails have almost no roughness and friction. For a transit vehicle to move forward on rails, it relies on its own multiton weight to impart friction to the rails. Transit movement on rails takes very little energy, but streetcars and trains cannot regularly climb slopes over 10 percent.[145]

Until the late nineteenth century, roads carried loads in the forms of walking feet, hooves, and wheels. Until 1888, wheels were all solid iron-, wood-, or rubber-rimmed. They were also narrow, under four inches wide, to reduce rolling friction. The weight of each wagon was concentrated on those four small rectangles of road contact at the bottom of each wheel. Where paving stones were wider than the wheels, each wheel would exert an uneven force on one side of the stone paver, thousands of times a day.[146] Millions of these forces would cause even the biggest boulder to crack, and crack again, until the road was an unruly mess of jagged and tilted rock. The expense and short life of these roads meant that most roads, most of the time, were barely paved at all.

The costs of cobblestone and asphalt were comparable, but the maintenance of a smooth sheet of asphalt was less incremental than a cobblestone street. Adjacent landowners or street crews could replace broken or unset cobblestones one at a time as they wore out. Replaced one cobble at a time, a cobblestone street still looked like a cobblestone street over the years. You could not repave a square foot of asphalt here and there every year and expect anything but an uneven mess after a decade or two. Cities had to take responsibility for paving all the streets, instead of leaving it to property owners. Cities occasionally put bonds out to pave their streets, but were often disappointed by the cheap materials contractors used.[147] The state of Oregon first charged gas taxes in 1919 to pay for the work of paving what were formerly dirt and gravel roads.[148] By 1929,[149] every state had adopted this model of road financing. Those able to afford to move out to the new auto-suburbs were also advocates for paving roads in their suburbs and on the route to their jobs. The network of paved roads replacing old dirt farm and market roads grew quickly between 1900 and 1930. Roads were cheaper to build than rail lines, and vehicles were paid for privately, not by the government. The speed of the car kept even far-flung commutes to twenty to thirty minutes, as long as the roads were newly paved and the traffic wasn't too heavy, yet.

The bike, cheaper than a horse, moving more freely than a streetcar, and faster than both, was the first widely adopted form of private transportation. Like the car, it was a luxury at first, but by 1899, the bike had become a utilitarian part of everyday transport. In 1870 a bike cost half an average annual wage,[150] by 1885 a bike cost three months' average pay,[151] and by 1895 a bike cost six weeks' average pay.[152] In 1914 a Model T car could be bought for a third of the average annual wage,[153] and by 1935 the average car cost one fifth of the average annual wage—under three months' pay.[154] Both bike and car changed from luxury goods to necessities. The car became more necessary than the bike as we developed zoning laws to remake our regions to the scale of traffic, not transit, walking, or biking. The bike gave America the idea of independent transportation, but the hills and distances gave bikers the idea of motorized independent transportation.[155]

## The rise of the car in traffic

One of the first gasoline car makers in America was the same Albert Pope who invested so heavily in the Columbia Bicycle, with Hiram Maxim doing most of the design work.[156] The few steam cars that were built before 1880 were tinkerer's novelties or trackless steam locomotives. Most cars built after 1890 looked more like bikes than locomotives, especially the ones using quieter and more compact gasoline engines.

By the 1920s, even subway proposals were being replaced by the cheaper prospect of simply adding traffic lanes to busy thoroughfares. Adding new lanes for traffic was much cheaper than excavating, building, and operating complete subway systems, or adding tracks. After the stock market crash of 1929, financing for new transit projects fell away. Cincinnati started work on its subway in 1920, stopped with the 1929 stock market crash, and never resumed.[157] Only Philadelphia, New York, and Cleveland built any new subway lines between 1930 and 1970.[158] Highways, limitless lanes, and traffic were the new hope for America's transportation.

In the twenty years between 1890 and 1910, streetcars and interurbans were the everyman's way of getting around within and between cities, and cars were a rich man's toy. The first retail car, the Locomobile, was an electric car produced by the Duryea brothers. It retailed for $600 (the equivalent of $21,400 in 2020) in 1900,[159] an unheard-of sum in a time when the average price of a well-made buggy was only $20 (or the equivalent of $710 in 2020).[160] No working man had ever heard of paying so much for transportation. For the next ten years, car ownership remained the province of the rich and rural.

From their start, the car was widely despised as the toy of the rich.[161] Any mention of cars in conversation was tinged with class envy, and anyone who owned a car in the city was using it for sport and conspicuous consumption. It was a useful tool in the suburbs and farm country, as long as the pavement was good. The first mass-produced car was the Ford Model T of 1909. Introduced at $850 (the equivalent of $27,300 in 2010), its manufacture was optimized over the next fifteen years to reduce the price to $500 (the equivalent of $8,800 in 2010) by the time it was discontinued.[162] Ford wanted to make a car his workers could afford, not just the wealthy.

The used car market was growing, making car ownership affordable for many. Between 1910 and 1917, the growing used car market enabled a fleet of used car "Jitneys" serving as taxis or buses that competed with transit for riders along heavily traveled routes.[163] Ninety percent of American families had access to a car by 1925.[164] If a family didn't own a car themselves, they probably knew someone who did and could envision all the benefits of having one. Making cars for the masses and finding ways to pay for the roads and parking they demanded was a huge step on the way to America's mass motorization.

People had long since become accustomed to the inexorable and silent path of the trolleys and knew for centuries to give irritable horses a wide berth, but cars were something new. Powerful steam-, battery-, and gas-powered motors made cars faster than anything else on the road—much faster than the 10-mph bike, the 6-mph trolley, the 4-mph horse, or the 3-mph walker. Unlike the trolley, cars could travel anywhere, including residential streets where children had played for generations. The death of children in traffic accidents in the first two decades of the twentieth century was a disaster, causing car sales to dip in 1923.[165] When the urban poor or middle class heard about a friend's child getting hit by a car, the rage was not just about the car, but also about the class of their killer. Before 1920, the burden of proof when a car killed a bystander or biker was very much on the larger, heavier, and faster vehicle and its driver. This extended naturally from nineteenth-century jurisprudence, where the drivers of horse-drawn vehicles that killed someone were subject to arrest and charged with homicide.[166] Having a large vehicle in the walking and transit city of 1910 was a grave responsibility, with serious consequences.

Cars in traffic were safer than horses but more dangerous than transit. Between 1900 and 1905, the fatality rate per one hundred million travel miles for horse traffic was 1,100,[167] for car traffic, thirty-six,[168] and for trolley transit, eight.[169] Most of these deaths were nearby walkers, not passengers. In 1928 Philadelphia 75 percent of those killed by traffic were walkers, with a third of those younger than sixteen.[170] Traffic was too fast and took up too much space to be ignored, and it had changed from reviled toy of the rich to an everyman's necessity.

AUTOMOBILE TRAFFIC DEMANDED A RESPONSE, WITH THE THREAT OF ITS GREATER MASS, speed, and damage. Cincinnati considered the idea of speed governors, mostly because 25 mph was an injuring speed, whereas 35 mph could be a killing speed.[171] This was anathema to "motordom," the growing collection of car drivers. The ability to move faster than anything except airplanes was what gave cars their appeal. Traffic control changed from punitive in the 1910s to a form of utility management by the 1920s. By the '30s the idea was to give traffic all the lanes and parking it needed. Streets became "sewers for cars,"[172] an apt pejorative, as the road was now a mortal hazard to enter. Furthermore, the engineers who took over traffic management from the police in the 1920s saw their work as another utility management problem, not a public order problem. From 1850 through 1900, engineers designed utilities to solve problems that had stymied cities for millennia, with clean water, sewage, electricity, transit, and trash cleanup. Engineers came into their own as problem-solvers and designers of novel utilities in the service of the public good. And the free flow of traffic was becoming the public good.[173] Why shouldn't engineers be able to make just as much sense of traffic?[174] The fact that it took walkers, horses, trolleys, and cars different amounts of time to travel down each block made engineering street traffic a challenge. The easiest "solution" was to get parked cars, streetcars, walkers, and bikers out of the street wherever possible. And to make parking the problem of private property owners through emerging standards of zoning and subdivision regulations.

Since the 1920s, traffic fatalities have decreased in almost every way as road design and parking policy for the sake of traffic has become an established field of engineering (figure 2.8). The traffic fatality rate per Vehicle Miles Traveled (VMT) peaked before 1910 and has fallen consistently since then; the rate per road miles built peaked in the 1970s and has not fallen consistently. The more lane miles we build to address congestion, the more Americans are injured or die on our 8.5 million lane miles of road.

Biking and walking became subservient to the new engineered utility of traffic. Biking changed from an expression of freedom and liberation[176] in the 1890s to a child's toy by the 1930s. Biking became an activity limited to children and the truly daring. Walking became the undocumented act of the indigent. Between 1930 and 1970, the only worthwhile way of daily travel for most Americans was traffic. Nearly all the wealthier decision-makers were getting around in traffic in 1920, and they funded, planned, and built for more traffic.

The intermingling of walkers, bikers, transit, and traffic was a major cause of accident, injury, and fatality. If the unit of utility for the street was the number of people moved to where they wanted to go, then the car was an impediment to the utility of the street, because cars moved fewer people on more land than anything else; trolleys, bikers, and walkers all used less street space and

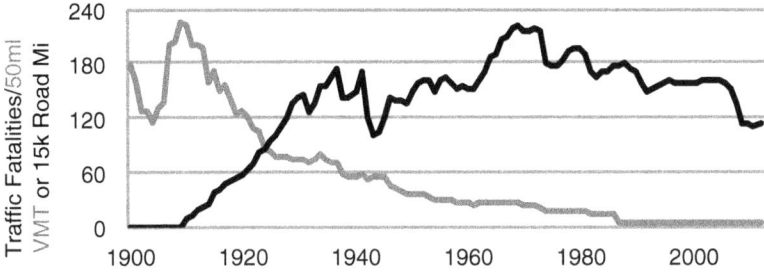

Figure 2.8. Traffic fatality rates by VMT and road miles, 1921–2010.[175] Note the significant decline in risk of miles driven, but the less substantial decline in risk of miles of road built.

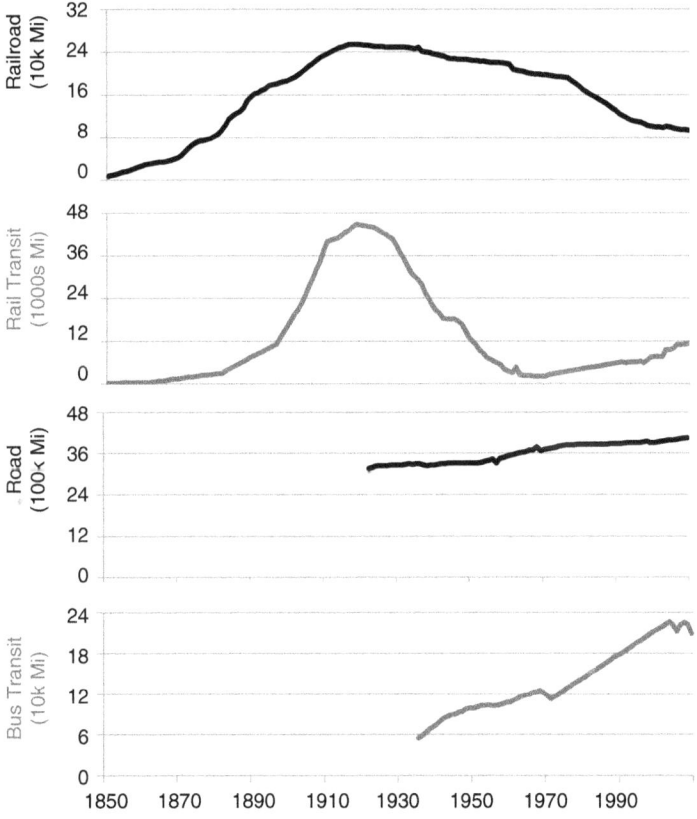

Figure 2.9. American history of transportation infrastructure, showing peak in rail and resurgence in rail transit, as well as growth in road and bus network miles. Scales differ by order of magnitude to fit comparison.[177]

urban land than cars. If the function of the street was getting people where they wanted to go in minimum time, then the car was superior to all else—but this assumed there was parking available at either end of every trip, that someone paid to maintain the roads, and that the vehicles were affordable to all.

Traffic dominates our transportation and land use policy. Before asking whether there are better alternatives, we must first understand the popularity and advantages of traffic in America. That's what we'll look at in the next chapter.

# 3

# Two Cheers for Traffic

**B**efore we consider what more or what better we might have, let's look at what we have already. Why should people consider alternatives to the car, unless they can deliver more service and value than traffic? Where people use biking, walking, and transit most, traffic is already of limited use, but those places are rare. In most places for most trips, traffic is the best way to get around. So, this chapter is about the many great benefits of traffic. It describes the myriad ways that traffic is obviously and inherently superior to any other mode of transportation. This is not to preempt the argument for transportation alternatives, but to understand what we have today, and where we can go from here.

## Traffic—freedom, access, and quality of life

While for some people the car in traffic is an unmitigated evil, to most Americans this is a weird and bewildering attitude. For most of us, traffic is as important and taken for granted as clean water from the faucet. Traffic is a quality-of-life issue for most Americans, except the tiny minority of urbanists, activists, and environmentalists who discount it as dangerous and inefficient. For most of us, being excluded from traffic is a disaster. Being taken away from our vehicles and from traffic strands us, preventing us from doing many of the things we want to. Only with traffic can we enjoy the full diversity and richness of homes, jobs, and play that is the American Dream. Your ability to purchase a car and participate in traffic means that you have made it. The best predictor of a family getting out of basic poverty is their acquisition of a car.[178] Affording and using a car is not only a signal of wealth, but a force multiplier for gaining more wealth and spending less time and money in getting the things we all need. For most Americans, traffic beats transit, walking, and biking in terms of comfort, convenience, and privacy.

In traffic, the car is yours. What you do with it is your business. It is your home and identity in traffic. You can listen to the radio, eat whatever food you want, and throw the wrappers in a little wastebasket or all over the floor, as long as you don't block your mirrors. You can talk on your phone, text discreetly at stop lights, brush your hair, tie your tie, loosen your belt, or whatever else you can do while sitting with one arm and your right leg reaching out in front of you. Many states have made some of those distractions illegal, but your car is still a little tub of privacy from the public, in public.

It is a simple matter to carry three or more passengers in a car, as long as all agree with the driver on time and destination. To move the same group or family on transit, you have to pay round-trip fares for each. If all of you were experienced and intrepid enough, you could ride bikes to your destination, but there'd be no rest, and a lot less conversation amongst your puffs and gasps. In traffic, you can carry several passengers free of charge, or hundreds of pounds of furniture as far as you care to drive, across town or across the country.

Cars and trucks can carry large goods and cargo on the same network as others carry their families. While regular cars, vans, and SUVs are six feet wide, and cargo vans, semi-trailers, and emergency response vehicles are eight feet wide, both types easily fit into the twelve-foot-wide lanes that connect all of America. Roads are built to carry a lot more than just cars. In America, 70 percent of the freight moved is by truck.[179] Nearly every grocery and department store in America is supplied by truck. Without the freight abilities of trucks and cars to and from stores, families would have to go to stores daily for all they could carry in their buggies and baskets.[180]

In traffic, you get to choose who you carry in your car. A car is even more private than your bathroom. Traffic and wind noise drown out whatever you sing, yell, or say to the rest of the world. In contrast, the people you share transit with are not of your choosing or maybe even your liking. Everything you do on bike or foot, you do in public. Only your car is a true refuge.

You can put as much or as little upkeep or improvement into your car as you wish. Many states require certain levels of upkeep, but none mandate the condition of the interior. That is your business, and you can keep it however you want. It is perfectly legal to keep your car full of clothes and old soda cups, as long as you can see out of the windows and reach the pedals.[181]

Your car is shelter in motion. No matter what season or weather, you can stay warm and dry so long as you have fuel. The military saw this advantage when planning for nuclear war in the 1950s. The interstate highways were billed as escape routes for a nuclear age,[182] and cars were emergency housing for the postnuclear world.[183] In uncongested traffic, you can move out of harm's way, or speed through areas you would not walk, bike, or wait for transit in. You can use a car any time of day, in the coldest hours of the night or the wettest hours of the day, to move wherever your restless heart wants. You don't spend a lot of your physical energy driving your car, which frees you to do whatever you'd like, so long as you pay attention to the road and keep control of your vehicle. To use traffic, you back out of your driveway and sit in the most comfortable chair you own to get practically anywhere you want in the country.

Another advantage of the privacy of traffic is that you can be in public without contacting anyone. This is particularly useful when you are sick with a contagious disease and need to get to the doctor. If you were on a bike or walking, you would be too exhausted in reaching the doctor for the visit to do any good. If you took public transit, you could touch or breathe on dozens of people, infecting them as well. The privacy and route efficiency of traffic lets you, and society, avoid these problems. Similarly, roads and traffic make it easier for emergency vehicles to get to you.

Federal and state governments have developed street standards to convey motor vehicles to the exclusion of all other modes for the last eighty to one hundred years. Regionally, this is through a hierarchical traffic system. The ideal traffic network is a hierarchy of local, collector, arterial, and freeway road types designed to make most traffic trips calm at the beginning and end, with danger and speed in the middle. Most land uses, like residences and businesses, are next to local roads, connected to arterials by collector roads. This avoids driveways connecting to larger roads with higher design speeds. Arterial roads add more and more lanes as they accommodate higher volumes of traffic and serve freeways of even higher volume at interchanges. Most trips in traffic, even the most trivial, move up through this hierarchy and back down, like the trip of water from roots to leaves through the trunk of a tree. Typical traffic journeys progress up from local to collector to arterial to cover large distances, and then back to collector and local road at their destination. Local roads may only connect to arterials via one or two collectors, and not to each other between subdivisions. Contrast the hierarchical road network with the grid or network street system, where few streets carry any more volume than the others. The hierarchical traffic system lets emergency vehicles move on wide highways to most houses, dealing with narrower streets only at the end of their sorties to reach the people in need. This is much faster for emergency vehicles than poking along multiple square blocks, with local, slow-moving traffic.

You control your route and your fate in a car, steering and braking in response to other cars on the road and in the parking lot. You are the master of your own destiny. You go where you want to go, when you want to go, how you want to go, making stops, and seeing sights as you want. You are not beholden to any bus schedule, taxi driver, or your weak legs. The nation is yours at the tap of your foot. The roads are designed for you as a driver to make traffic journeys as easy as possible. The roads are overtly not designed for anything but traffic.

Almost all trips in traffic carry you door to door. Free parking is almost guaranteed at the beginning and end of every journey. Developers cannot build and connect to the road system without adequate parking. If you want to build with less than adequate parking, you will need to explain several times to several levels of government why you have failed in your professional and civic duty to build enough parking. Inadequate parking is seen as a threat to the public order, as parking enables traffic to get out of the way as soon as it gets where it is going. Only in the busiest or oldest places are you going to have to search for or even pay for parking. To complete every traffic journey, we require parking spaces at every beginning and end. Most are freely provided for our parking convenience by the developer, builder, and shop owner as a cost of doing business for customers. In traffic, you are welcome nearly everywhere.

Because anything worth traveling to is already on the road network, changing your housing or workplace is not a problem. If your job changes locations, you just start a different daily route and routine in traffic. In contrast, transit limits you to those places served by transit. If you are dependent on transit, you can lose your job if it relocates too far away from a transit route—you are effectively laid off. This is a compelling personal case for getting a car as soon as possible and keeping it as long as possible. The most reliable tool for moving from the lower class to the middle class in America is getting a car, as both work and routine trips become less time-consuming and more reliable for most Americans in traffic. Traffic liberates the poor from the schedules of transit, the slowness of biking, and the limitations of walking.[184]

Because of this hierarchical traffic network and plentiful parking at nearly every origin and destination, most traffic trips begin and end in peace, even if the middle of the trip involves jousting in heavy congestion and crossroads. Local roads in subdivisions are never meant to carry any more traffic than needed for the houses or businesses on that street.

America has an enormous network of roads and parking. In the nineteenth century, long-distance and local roads were dirt, gravel, cobblestone, brick, or wood that wore out quickly and turned to muck within a few seasons. In the twentieth century, America doubled its miles of road and paved almost all 4.5 million miles of them. Paved roads are more durable, professional, and expensive than these ephemeral materials of our ancestors. America has two billion parking spaces for 280 million registered vehicles.[185] This much parking was required by zoning and building codes that developed at the same time as traffic, to serve the developing needs of traffic.

We can now reach places our founders could not have imagined. They could not have seen Nebraska, Devil's Tower, or the Rocky Mountains without months of hazardous walking on disused paths. Two hundred years ago, the trip between Boston and Washington, DC, took as long as a traffic trip across the country does today, with a much lower chance of survival.

We can get to anywhere that has a road, and we have gotten very good at getting roads everywhere. Counties and whole states that once relied on their river ports and rail depots for trade can now subdivide and sell all the land they don't care to till. Every part of America is accessible on a road and parking network that takes up less than 2 percent of the land. From curb to curb and in every parking lot, the practice of civil engineering has evolved to create obvious paths for cars to easily and thoughtlessly flow from origins to destinations. Everything is standardized to maximize fulfillment of driver expectation and minimize the risk of a collision or congestion. An area equal to West Virginia is now paved for cars—about half for roads and half for parking.

The efficiency of a car's motor is astonishing. Consider that a three-thousand-pound vehicle that gets thirty miles per gallon is traveling 158,400 feet on 0.13 cubic feet of fuel. The stream of gas that the car runs on is not much thicker than a human hair.

The most dangerous kind of traffic is mixed traffic, where multiple modes move with different speeds, masses, and directions. This was traffic in America before 1930, and this still is traffic today in much of the developing world. The notion of complete separation of walkers from traffic developed in the 1920s as a reaction to the chaos of mixed traffic. The more familiar car ownership is for a nation, the safer traffic becomes (see figures 3.1 and 3.2).

The earliest traffic planners attempted to impose order on the chaos of mixed traffic, with the added danger of the speed and mass of the car.[187] First they proposed rules that few obeyed, then tried punitive policing, and finally engineered everything but motor vehicles out of the trafficway. Civil and traffic engineering became the art of separating modes of different speeds from each other. The roadway has been designed to be the sole province of traffic ever since 1930. Traffic fatality rates of walkers and slower horse-drawn carriages peaked in the 1930s, until they were excluded from the trafficway. Today's traffic is much safer than ever before, even for bikers and walkers. Of course, some of that safety is the result of most not even bothering to bike or walk.

Zoning and traffic engineering arose a century ago to get anything other than a car off the streets. Traffic has been getting safer since we got rid of congestion of mixed modes sharing space in the streets, put in standardized signals and road markings to make driving nearly mindless, and made seatbelts mandatory. In the forty years since the seatbelt, we've made traffic ever safer with continued advances in roadway design and intelligent transportation systems.

A couple of centuries ago, right of way (the strip of land deeded to the state for the use of a highway) was the sole province of highways of national importance, like the National Road or the Boston Post Road. Urban streets were just the space between buildings. Paving was often the responsibility of adjacent landowners.[189] When states and cities took over the regulation of fast, heavy,

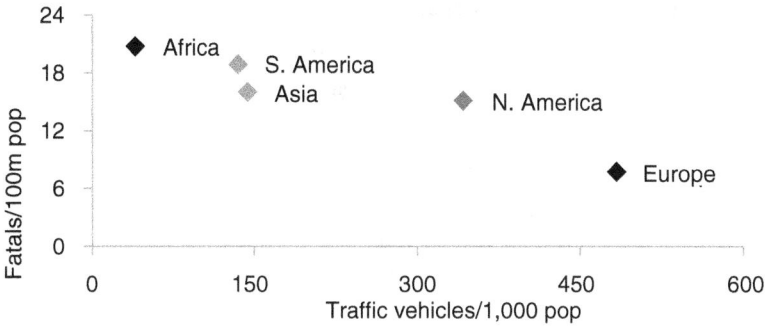

Figure 3.1. **Car ownership vs. traffic danger by continent.**[186] **Note that the US is only part of North America.**

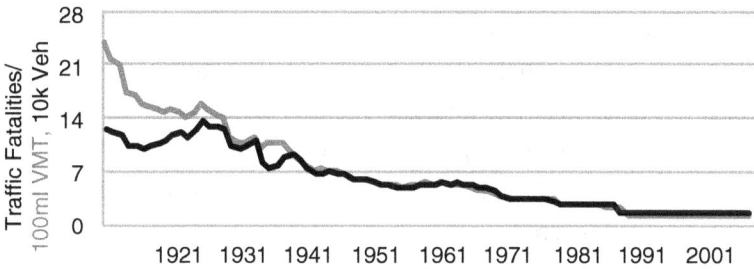

Figure 3.2. **Improving safety in traffic, showing declining fatality rates in traffic relative to VMT, and vehicles in traffic.**[188]

and numerous cars in traffic, right of way became ubiquitous. The Bureau of Public Roads set up the financing and right of way standards that directed gas tax revenues to states. This financed waves of roadbuilding in state roads during the 1920s, the national roads in the 1930s, and the interstate highways in the 1950s and 1960s.

# The costs of traffic

On some days, you may not spend a penny to operate your car. On other days, like when you need fuel or a brake job, your costs are substantial. Even though transit is cheaper for its users than traffic, it is a daily expense, whereas the only times car owners incur expenses are at the pump or repair shop. Transit users feel their costs every day, unless they use a weekly or monthly pass; traffic users deal with costs only when they fill up gas, for sporadic repairs, and taxes.

Gas is cheap in America, especially compared with Europe and Asia. The US gas tax accounts for between 2 and 4 percent of the cost of car ownership, depending on diverse state policies and the variable price of gas. Because we have such a large installed base of highways, we don't need high gas taxes to cover new road building. By 2005, we spent as much on highway maintenance as expansion, though the costs have shifted even more to maintenance since then (figure 3.3).

Apart from our homes, the car is the largest expense for Americans, and we shell out money for new cars several times over our lives. You spend so much on it, you'd be a fool not to use it to go everywhere.

Your car is not the government's. The government does not pay for it. Aside from occasional inspections, the government does not invest in the upkeep of your car. Federal, state, and local governments design, construct, and maintain the millions of miles of right of way and roads that unite us. They also pay to enforce traffic laws. Over 15 percent of law enforcement budgets go to traffic law enforcement (figure 3.4). Our police and highway departments are as good at cleaning up traffic accidents and restoring the "free" flow of traffic as grocery stores are at cleaning up broken glass on aisle six. As the figure below shows, traffic enforcement budgets rise and fall generally with traffic fatality numbers.

In contrast, with transit, everything is the transit agency's problem: the tracks, stations, vehicles, and policing. Passengers notice immediately if maintenance is deferred or schedules slide. If transit is not policed and cleaned every hour, its deficiencies are seen, felt, and smelled by its passengers daily. Graffiti, derailments, and terrorism are among the manifold worries of transit agencies. If something goes wrong with your car, it is your problem. If something goes wrong with your bus, it is their problem. It is much easier to blame problems with transit on specific people, routes, and agencies than it is to blame problems with traffic on millions of drivers and miles of roads.

For the last century the way governments have paid for most of the roads was with the gas tax. This is an elegant mechanism, tying road use to construction, expansion, and maintenance. The more people use the roads, the more gas they consume, and the more money there is for more roads. This positive

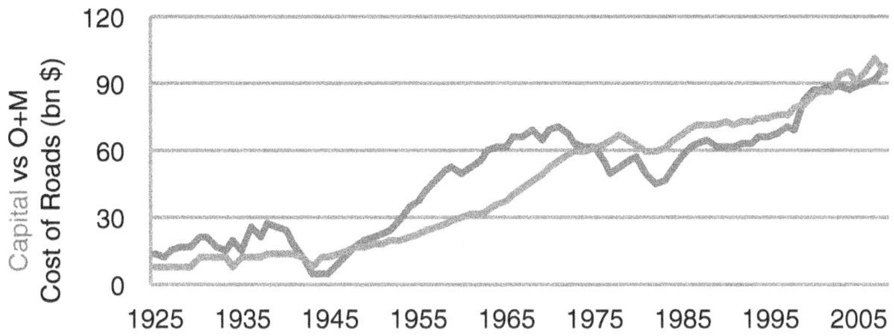

Figure 3.3. Capital versus Operation and Maintenance (O&M) costs for the roadway network.[190]

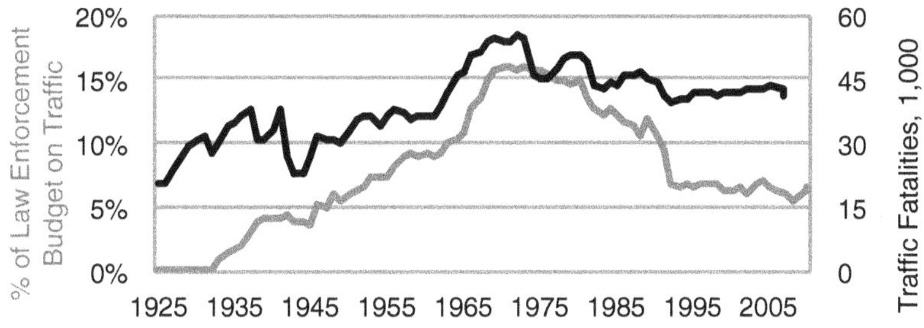

Figure 3.4. Law enforcement devoted to traffic versus traffic danger, showing fluctuation with overall traffic fatalities.[191]

feedback loop has resulted in miraculous expansion of the road network in the twentieth century. Of course, the largest user fee in traffic is in the purchase, upkeep, and time spent driving the vehicles themselves. The government is well-served by traffic. They just have to build and maintain the roads, and the public will pay for and operate the vehicles to get wherever they are going. If businesses or buildings want to connect to traffic, they have to give up some of the land for parking. Traffic is funded mostly by user fees and exactions from millions of drivers, developers, and taxpayers, whether they drive or not. Traffic is not only a democratic means to let the most people get where they want, but also a great savings for the government.

Governments also pay to keep roads dry. One quality of roads, parking, and shelter is the ability of their impervious coatings to shed water and return to dryness as soon as possible after each rainfall. Impervious and crowned roads jettison water quickly to keep us safe even in torrential downpours.

## Other traffic benefits

Nature was once a remote, unreachable thing. The largest parcels of nature away from the influence of a road are in Montana and Alaska, where the cost of developing land still outstrips the return from its use. Traffic lets us see the edges of nature all along our 4.5-million-mile road network (figure 3.5). This is a luxury not enjoyed by most bikers, walkers, or transit users. The most remote transit station will take you to a parking lot in the exurbs or near an airport. Walking to untrammeled nature from many places would take days. Or you can just hop in traffic and be next to some nature with the tap of your gas pedal in an hour or two.

Roads have been a boon to development. Without good roads and fast transportation, development was limited by how much food and materials we could get to a place. Now that half of the nation is within a quarter mile of a road (see figure 3.5), almost any land can be farm, factory, or subdivision. Roads let land respond to market opportunities like never before.

Roads have been a boon to agriculture as well. From the start of the Union, the point of good roads has been moving food to markets. They let farmers get their goods to market, silo, or warehouse no matter the weather. The Midwest was covered in water much of the year before wide-scale wetland drainage in the 1870s.[193] Drainage opened the Midwest's thick soils to grow more grain than the US has needed ever since. Iowa soils were adapted to water and became impassable mud with every rainfall. Thomas MacDonald, chief of the Bureau of Public Roads for the first half of the twentieth century, came out of this seasonal immobility to plan and build a road system on top of these soils, no matter what the climate, geology, or soil type. Roads that were impassable muck every spring through the 1920s were rebuilt as professionally designed, paved, and dry highways by the 1950s, and have remained so ever since (figure 3.6).

The car has advantages of ownership, route, capacity, financing, infrastructure, power, and safety over any other transportation mode. No wonder it is the signal possession of the comfortably middle-class around the world. It is almost a given that the first thing households will do with their first taste of wealth above poverty wages is buy a car. Only with a car can they reach the even greater opportunities open only to the complete traffic network.

Figure 3.5.  All 4.5 million miles of roads in the continental United States, 2010.[192] Due to resolution issues, the roads could not be drawn to scale, but they are a paved area greater than Maryland and half the area of Kentucky.

Figure 3.6. Dumfries Road in Prince William County, VA, as a dirt road in 1919, a single paved road ca. 1930, and as a four-lane arterial in 1989.[194] (All three views courtesy of Virginia Department of Transportation, collection of Tom Saunders.)

Now that we've described some of those advantages, in the next chapter we explore the *dis*advantages of traffic, before moving on to consider possible alternatives, and how to diversify America's transportation options at the same or lower cost, with greater prosperity and opportunity.

# 4

# Problems with Traffic

The following chapter talks about some of the costs of traffic to America, including safety costs, economic costs, and livability costs. This is a counterpoint to the previous chapter, which praised the benefits of traffic. For this book to make the argument for an alternative to traffic and the restoration of choice to America's transportation, we need to consider both the good and the bad of traffic dependence.

## Managing traffic for congestion

Looking at pictures or film of American streets from the first decade of the 1900s shows that the streets were not so much crowded as mixed. Even streetcars were moving at the speed of a person walking or a horse walking. When automobiles became common and then dominant in the streets, they demanded ever more clearance to avoid congestion. Congestion has been a common complaint of traffic users for over a century. Congestion—slow traffic— is also one of the safest conditions for traffic, other modes of transportation, and surrounding walkers.

Roads are engineered to serve "driver expectation." Because drivers are piloting vehicles with greater momentum than bullets,[195] the mission of the traffic engineer is to sign and mark the roadway into obvious paths for traffic. This means that lanes, signs, sight distances, and signals must all be standard and predictable. Things like buildings and trees near the road are "hazards" to be eliminated from the "clear zone." The aim of all this is to make roads that can be used with a minimum of thought from drivers. Ironically, most fatal collisions happen on the clearest roads designed to require the minimum diligence from drivers. "Well-designed" roads are overbuilt for most of the day's traffic, allowing careless speeding most of the day. Speed is what kills if anything goes wrong. It is hard to kill yourself in a traffic jam, even if you want to die from the tedium.

Table 4.1. The nine most congested and dangerous stretches of road in the United States. Note that these are two different lists.[196]

| Rank | Road Name | Location | Rank | Road Name | Location |
|---|---|---|---|---|---|
| 1 | I-95 | NYC | 1 | Dalton Hwy. | Alaska |
| 2 | I-405 | Los Angeles | 2 | CO 550 | Silverton, CO |
| 3 | Van Wyck Expressway | NYC | 3 | US 19 | FL |
| 4 | I-10 | Los Angeles | 4 | US 2 | MT |
| 5 | CA 91 | Riverside, CA | 5 | US 129 | NC |
| 6 | Long Island Expressway | NYC | 6 | I-15 | CA to NV |
| 7 | Brooklyn-Queens Expressway | NYC | 7 | I-26 | SC |
| 8 | I-90 | Chicago | 8 | I-95 | FL |
| 9 | I-5 | Los Angeles | 9 | CA 101 | Los Angeles |

The more popular and prosperous a place, the more people will want to go to it. If traffic is the only way available for people to get to it, there will be congestion. The most congested roads usually connect the most popular places. Look for the most legendary traffic congestion (table 4.1)—such as the Van Wyck, Long Island, or Brooklyn-Queens "expressways" in New York, or the 405 in Los Angeles—and you will find the most prosperous job and housing markets in the area. Conversely, where land use is minimal, or the economy is declining, congestion is much less of a problem on the roads. What is the point of traffic congestion when no one wants to get to anything?

Congestion is usually caused by the last 5 percent of traffic to get on the road. The more congested traffic gets, the more new drivers consider another way to get where they are going. Past a certain point of congestion, no new vehicles will enter the stream of traffic. The next 5 percent of traffic that would have brought the congestion to a standstill will find another way around the intolerable traffic jam or just wait it out at home or work. As Yogi Berra would

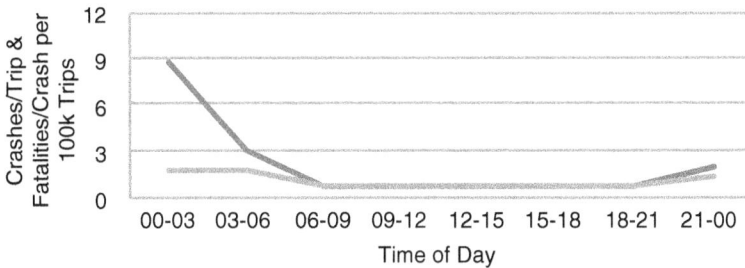

Figure 4.1. Traffic fatalities per trip by time of day. The most congested times of day are also the safest.[198]

say, "No one goes there anymore, it's too crowded."[197] People in congestion yearn for two things: to get where everyone else is going and that fewer people would go there.

One upside of congestion is that it is the safest time to drive (figure 4.1). If no cars are getting up to a killing speed, they cannot do serious damage to themselves, other cars, or bystanders. A long line of traffic moving at the same speed as a stagecoach may be frustrating for those in it, but it is safer than an open road.

Crashes at night, whether between vehicles or by running off the road, are more severe due to less congestion and more speed upon impact, resulting in more fatalities. The distribution of traffic is skewed toward the daytime, "waking hours," nationally, while fatalities are spread more evenly around the clock. The same pattern holds even if weekends are excluded, eliminating the majority of alcohol-related crashes.

If congestion saves lives, it is ironic that the first goal of traffic engineering is reducing congestion. The reason is that congestion is the major cause of driver stress in traffic,[199] not high speeds or the chance of getting into a wreck. Drivers do not complain of roads that are too fast, and complaints over severe crashes on main stretches of roadway are intermittent, and are usually seen as the fault of the driver, not the road design. Most often, drivers complain when their route is consistently congested. Highway engineers making design decisions and lawmakers in control of budgets have to listen to congestion complaints more often than complaints about safety. No wonder we build new expensive roadways, bypasses, and connnectors as though congestion is the major problem with traffic.

# Law, death, and the right of way

Over the last century, America has changed its laws to define the trafficway as an industrial conveyor, a sewer, for the sole purpose of conveying traffic. Without reciting the history of how we lost our rights to the streets, this section describes the current state of traffic jurisprudence in America, and how tenuous our rights are in the face of traffic.

Before I got my driver's license, I biked everywhere in Atlanta. For two years I worked second shift at a bookstore, biking on the edge of the road five miles to work and five miles back at midnight. There were no bike lanes in the whole of Atlanta back then. The peril of this mode was not lost on me, but I biked anyway. I never got hit, and traffic went around me when it could. I'm writing this for you because I moved on from that commute.

Nowadays, I bike to the coffee shop to write, two miles there and two miles back, on the barely used sidewalks. I live twenty miles away from the focal city now, in the suburbs. I know that the six-lane highway has a 35-mph official speed limit, but a design speed of 45–50 mph in free-flowing traffic. I would be killed on this road as easily as a driver adjusts their radio or scolds their passenger. The nearest bike shop has a significant amount of retail space devoted to helmets, as I'm sure "getting hit by a car" is top of mind on these suburban streets. The further away we get from historic city centers, the more that places and the roads that connect them are built out to maximize speedy throughput of traffic.

In 1976, John Forester, an early modern biking advocate, supported a muscular and educated brand of cycling called "vehicular cycling."[200] Forester claimed that bikers should not creep apologetically along in the gutter but claim the entire lane, as they had the same rights and responsibilities as cars and trucks in traffic. Vehicular cycling was safe as long as the biker was able to move and accelerate as a full member of traffic. You had to start good and get great to bike amongst traffic. All bikers had to be as fit as Ferraris. Vehicular cycling makes more sense where bikers are an expected aspect of traffic, and not a bizarre exception to suburban "rights of way." Part of Forester's motivation was the terrible injustice visited upon cyclists who were killed in and by traffic. Car drivers often never saw a day in court, and the families of dead bikers were left to mourn without recompense. In many cases, car owners sued bikers that they hit, for the damage to their cars.[201]

**ANOTHER CASE THAT ILLUSTRATES HOW THE LAW VIEWS TRAFFIC AND ITS VICTIMS IS** that of Aimee Michael. She was driving her BMW on Camp Creek Parkway in Atlanta to pick up cake and ice cream for her family's Easter lunch on the afternoon of Easter, 2009. On a curve, she sideswiped a Mercedes SUV occupied by the Carter family. The SUV crossed the median and hit a Volkswagen Beetle driven by Tracie Johnson. The SUV caught fire, killing all four family members inside. Tracie Johnson was severely injured, and her daughter was killed.[202] Mrs. Johnson's husband and family, in a vehicle in front of the Beetle, witnessed everything in horror through their rearview mirror.[203] Ms. Michael slowed down for a bit, saw what she had done, then sped off.

By fleeing, she was damned. The horrific accident made the news, and the hunt was on for the villain who had killed five people without the grace to stay around for the police to answer questions and trade insurance information with the bereaved. Within a week, the police had tracked down Ms. Michael, as she was trying to get bodywork done on her BMW to remove evidence of the collision. They caught her just as the Carter and Johnson families were done attending memorial services.

In trial, the judge showed no leniency toward Ms. Michael and her mother for fleeing the scene of a deadly collision and trying to hide the evidence. "The one thing I couldn't get out of my mind is that you left the scene of the crime," Fulton County superior court judge Kimberly Adams told Ms. Michael: "And then you followed a course of conduct designed to cover up what happened." A state trooper interviewed about the case said, "If she had not fled the scene, she probably would have gotten a misdemeanor charge for an improper lane change, but since she fled, the charges were increased to vehicular manslaughter." She would have been charged with improper lane changing because that was all the wrong she did in the eyes of traffic laws. The assumption was that the Carters' and Johnsons' lives were forfeit upon entering the roadway. Unless the car that kills you is witnessed driving egregiously, the driver will be charged with bare minimum charges. Aimee Michael got a thirty-six-year sentence for five counts of vehicular manslaughter. Not because she accidentally killed five people, but because she fled the scene and tried to cover it up.[204]

Raquel Nelson, also from my hometown of Atlanta, offers another example of the meaning of roads and traffic in America. On the evening of April 10, 2011, she and her three children got off the bus at a stop on Austell Road directly across from their apartment complex. They had missed a transfer and had had to wait ninety extra minutes for this bus. It was after dark when they finally got off at their bus stop, across a five-lane highway from their home.[205] The closest crosswalk with a signal was over a quarter mile up the road, a fifteen-minute walk for adults, longer for a mother with three kids.

Several residents from Ms. Nelson's apartment complex were at the same stop and in the same situation. They all did the most obvious thing. They crossed the twenty-four feet of road toward the median left-turn lane, where there was no traffic, and looked out for the headlights of cars to know when they could cross the last twenty-four feet to safety and home. It was night, so they had to pay attention to the headlights and judge their oncoming speed to know when it was safe to cross. Another resident made a break for it and got across.[206] Ms. Nelson's youngest son, age four, took that as a cue that it was safe to cross, so he did. Ms. Nelson ran out after him and called for him to stop. That's when they were hit by Jerry Guy, a quarter-blind man who had been drinking and taking painkillers before driving that night. The average speed on that part of the highway was over 50 mph. Ms. Nelson and her daughter were grazed. Her son was killed. Mr. Guy probably wondered what that "whump" was, as he drove at the same speed as every other car on the highway that night.

Mr. Guy spent six months in jail for a hit-and-run he may not have been aware of.[207] Ms. Nelson was sentenced to three years for vehicular homicide for allowing her son to cross the road. The road is for cars. The only place a walker has any legal standing is the crosswalk, and only if they have the green light. That is all they get. Anywhere else, their lives are forfeit.

The prescriptions and parables around this case have ranged from the responsibility of the transit agencies to site their stops near crosswalks to the responsibilities of walkers to not cross the road except at crosswalks. But the right of traffic to move as fast and freely as possible is never questioned.

The law did not always favor the car driver. When automotive traffic was a new element of our towns and cities built for walking, horse, and transit, drivers were presumed to be at fault in collisions. The evening of Wednesday,

September 13, 1899, real estate broker Henry Bliss stepped off a trolley in the Upper West Side of New York City into the middle of the street one block from his house and two blocks from where John Lennon was shot eighty-one years later.[208] He turned to help a woman he knew down from the trolley. He took his attention off the street and was hit by an electric taxi driven by Arthur Smith. The electric car knocked him to the street and ran over his head and chest with both of its right-side wheels.[209] The taxi's passenger was a medical doctor, who rendered aid and comfort to the gravely injured Bliss and called for an ambulance. Bliss was taken quickly to Roosevelt Hospital, where he died the next morning. He may not have been the first American killed by a car in traffic,[210] but he was the most well-known.

Mr. Smith was charged with manslaughter, and his famous passenger was somewhat implicated in the scandal. Doctor David Edson was the son of the former mayor Franklin Edson[211] and some had suspected him of prompting undue speed from Smith. Smith may have been moving around a delivery truck that was parked on the right side of the road, but neither could see the people alighting from the trolley and their taxi was moving too close to the trolley to be safe. Smith was acquitted as the killing was unintentional, and the furor over Edson quickly subsided.

In the 1920s attitudes changed, and the driver was no longer automatically suspected. Drivers were viewed less negatively as the car changed from being a rarely-encountered rich man's extravagant hazard to an everyman's instrument of freedom. By the end of the Depression and the War in the Pacific, the car was the undisputed engine of American freedom and opportunity.

FOR THE LAST EIGHTY YEARS, THE BURDEN OF PROOF FOR TRAFFIC ACCIDENTS HAS BEEN on the struck, and not on the striker.[212] Through the 1910s, cars were seen as the province of the rich and careless, a perfect proxy for class resentment during the Gilded Age and the union wars. The first uses of the term "jay-walker" between 1910 and 1915 indicated a shift in the attitude toward cars. Cars were changing from exotic luxuries to hazardous everyday realities. Only a country "jay" would not be wary of traffic's mass and speed on the city streets. The 1920s were important years in the relationship of cars to the street. In 1922 cities were holding memorial parades for the hundreds of children killed by traffic every year. Safety councils were publishing lurid posters warning of death at the bumpers of traffic. In 1925 Los Angeles published their traffic standards, limiting the movement of walkers to crosswalks at signals, removing their former right to the entire street. By 1930, traffic was no longer a monster or a Moloch, it was a beneficial and universal force to be heeded by good children. Only bad children would play in the street. And the street would mete out their just deserts as soon as morally possible.

Railways established the precedent for traffic's deadly "right of way" eighty years before the rise of automotive traffic. Railways have long claimed property rights and control of the area around their tracks. Ever since we started moving tons of coal, freight, or passengers with heavy vehicles, the danger has been obvious. Everyone around these enormous wagons or massive railcars was advised to stay away from them. Their deaths would be inevitable if they fell under the wheels or were struck by the cars in motion. The momentum of railways was never the conductor's fault, as it took a mile or more to stop a loaded train. People on the train tracks were considered trespassers, and their deaths were considered to be entirely their fault.

The deadliness of the roadway was simply inherited from the deadliness of the railway.

## The inherent danger of traffic

Traffic—the independent piloting of millions of vehicles on America's streets, roads, and highways at high speeds—is an inherently dangerous transportation mode. After a decrease in traffic volumes and switch to lower-cost modes between 2008 and 2011, traffic has rebounded in volume and danger. Traffic fatalities are now as numerous as they were in 2006, over forty-two thousand Americans per year, partially because less traffic is traveling on roads overbuilt for peak volumes from a decade ago.[213] The rate of traffic crashes, injuries, and fatalities is no longer as concentrated in the night and midday off-peak.[214] Traffic is the leading cause of death for Americans between the ages of ten and thirty-five, and accounts for a third of the deaths of high school- and college-aged kids (figure 4.2).

With each successive wave of highway development and paving for traffic—from the cities in the 1900s and 1910s, to the states in the 1920s, to the

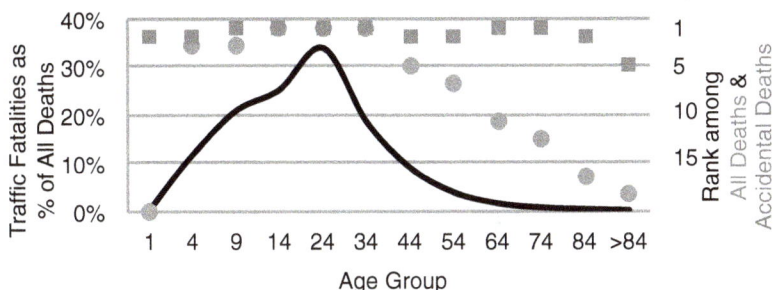

Figure 4.2. Traffic is the first or second leading accidental cause of death for all age groups in America until after age seventy-five (dark gray squares read on the right axis). Traffic is among the top three causes of death until age thirty-five (gray squares read on the right axis). Traffic causes a third of all deaths for twenty-five-year-olds (black line read on the left axis).[215]

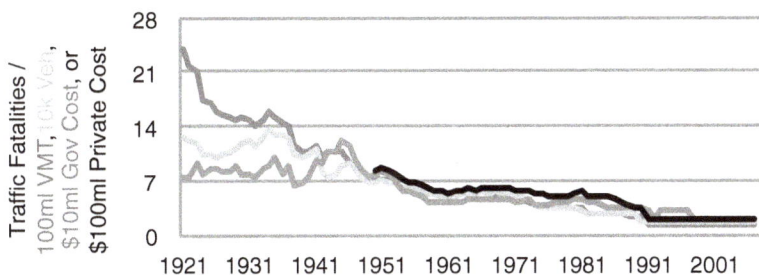

Figure 4.3. Improving safety in traffic, showing falling chances of being killed by traffic per mile driven and population.

US highway system in the 1930s and the interstate highways in the 1950s and 1960s—speeds and hazards increased. The peak death tolls from traffic were twenty-four fatalities per one hundred million VMT in 1921,[216] 294 per million population in 1937,[217] or 56,000 deaths overall in 1972 (figure 4.3). American traffic death rates were at their lowest in 2014, still claiming 32,744 American lives in that year. Since the pandemic quarantines of 2020, traffic fatality rates have shot back up to 2007 levels, with 42,915 Americans killed in traffic in 2021.[218]

Something had to be done about the "dieways."[219] Then, as now, the largest group of fatality accidents were high-speed, single-car accidents where vehicles left the roadway and hit something beyond the roadway. In response,

Congress passed, and President Johnson signed the National Traffic and Motor Vehicle Safety Act and Highway Safety Acts in 1966, which gave traffic and mechanical engineers greater responsibility for safety of the roads and the vehicles upon them.[220] For roads, engineers prescribed the obvious solution: a clear zone around every road. This clear zone would allow drivers the time to notice that they had left the road and steer their cars back onto the road or at least diminish the severity of their collision. The clear zone was cleared of all things that could stop an out-of-control car, but people could walk there, because they were less of a hazard to the wayward car's progress. While the impetus for the highway safety acts was the rise in freeway fatalities, engineers applied the clear zone rules to every new road, no matter how urban. Walkers and bikers got the clear signal they were not welcome in the new landscape.

Highway and traffic engineers make the roadways very predictable to minimize this inherent danger. We all have to pilot our vehicles on these roadways and have to be ever-vigilant to avoid death. We are always thinking about collision, damages, and even death while driving, even if we have never been in a crash, otherwise we would drive much closer to each other in traffic. We all know the stakes of colliding with another car in traffic, even if we don't know the specifics. The only attitude traffic warrants is constant vigilance. It is easy to die in traffic.

## The space cost of traffic

America has built the rules of the road over that last century assuming that people are going to move in traffic at a certain rate of speed between things that are spaced far enough apart by zoning and parking to be usable by traffic. When people try to navigate this landscape out of traffic, they are inconvenienced, discouraged, and endangered. The following section is about just how much traffic requires America's landscape to be expanded until it is only usable by traffic.

Traffic, like biking, walking and transit, flows best where its lanes are smooth and predictable, usually paved in asphalt or concrete. Traffic's space needs per vehicle and per passenger are much higher than the other three modes of transportation. There are several problems with this, including the heat-island effect of all that green space converted to paved space, and the runoff from all that paved surface. Paving urban and suburban landscapes for roadways and parking lots also means that water that formerly percolated down into the local soil is directed to the nearest stream and away to the ocean. Most urban streams resemble desert streams, with deep, scoured banks. The scouring is from the sudden flows of stormwater jettisoned from all the paved area after even small rain showers, as well as the lowered water table for base flow. Modern stormwater management has been evolving since the 1960s, but America's paved landscape has been growing since the 1920s. The damage has been done, and the stormwater still needs to be dealt with for every new paved acre.

**CONSIDER A TYPICAL, MODERN CLOVERLEAF INTERCHANGE IN THE SUBURBS OF** Washington, DC. An eight-lane east-west interstate highway carries 272,000 passengers per day (twenty-three thousand per lane per day), while the traffic carried on the six lanes of the north-south road is fifty-nine thousand passengers per day (about ten thousand per lane per day).[221] The rail transit line down the center of the freeway carries twenty-seven thousand passengers per day (13,500 per track),[222] but here it is at the end of the transit line[223]—further east, the transit line carries 144,000 passengers a day, while the highway carries 130,000 passengers (figure 4.4).[224]

A typical, modern cloverleaf interchange, with highways, collectors, and ramps, requires ten paved acres within thirty-three acres of land. This footprint is the same as or bigger than St. Peter's Basilica in Rome, the Burj Khalifa in Dubai, a supertanker, a typical American shopping mall, the Khufu pyramid at Giza in Egypt, the Empire State Building, Devil's Tower in Wyoming, or the Eiffel Tower in Paris. We devote so much space to traffic to convey so many cars and trucks safely to where they are going.

**TRAFFIC MAKES UGLY, UNUSABLE LANDSCAPES FOR EVERYTHING BUT TRAFFIC, BUT THAT** has only reinforced the urgency for most Americans to get a vehicle so they can be in traffic. Most Americans love traffic and have no problem with its expansion. Beautiful buildings and pedestrian scale are nice, but not nearly as nice as easy access to jobs, housing, and shopping throughout regions built dozens of miles across in the last century. The road space taken up by a car in

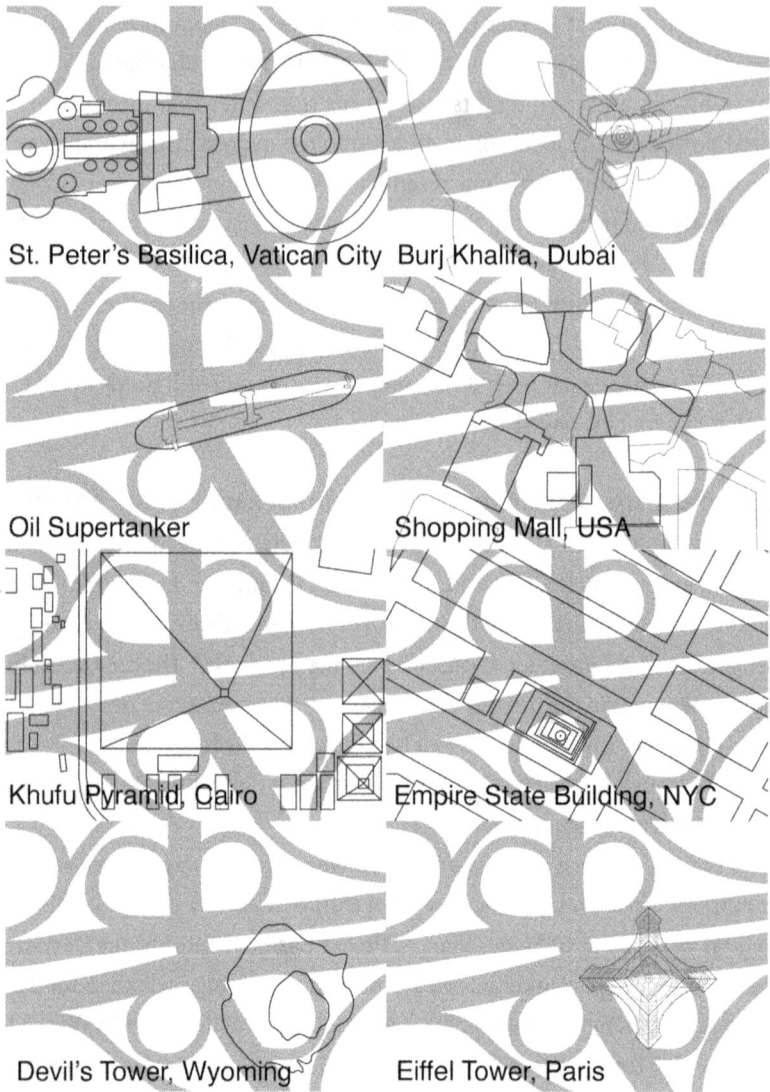

St. Peter's Basilica, Vatican City   Burj Khalifa, Dubai

Oil Supertanker   Shopping Mall, USA

Khufu Pyramid, Cairo   Empire State Building, NYC

Devil's Tower, Wyoming   Eiffel Tower, Paris

**Figure 4.4. The area of a typical highway cloverleaf compared with some very big things.**

traffic depends on your speed and the lane width. You need at least two seconds of safe stopping distance to react to the actions and mishaps of the vehicle in front of you. At the national average speed of 31 mph, that is ninety-one feet, or almost three dashes on the road. Cars in traffic may consume these dashes like Pac-Man consumes dots, but they actually represent a lot

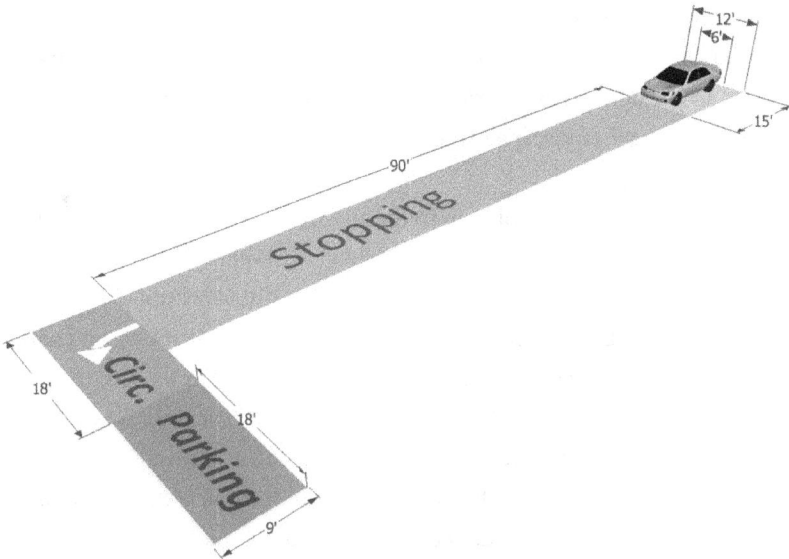

**Figure 4.5.** Space needs of the average car in traffic including twelve feet width for a driving lane, ninety-one feet for stopping distance, and 325 sq. ft. for turning into and stabling in a parking space.

of space. Each of the dashes on the road is forty feet apart. For scale, consider that your home is probably less than forty feet wide, and may be less than forty feet deep. The average width of a lane is just about twelve feet, so each car in traffic needs 795 sq. ft., or almost 2 percent of an acre to move through wherever they are driving. Christopher Columbus's flagship, the Santa Maria,[225] would have fit easily in two dashes and two lanes of today's traffic. Add space for parking, and you need 1,900 sq. ft. per car (figure 4.5) or almost 3 percent of an acre per car.

The landscape built out for traffic is most usable by traffic. Bikers and walkers are relegated to the margins of the rights of way, to journey toward destinations made distant by miles of roads in a hierarchical network designed for traffic trips, not walking and biking trips. Once walkers and bikers reach their destinations in American landscapes built for traffic, they have to traverse the inevitable parking lots offered to drivers in traffic along every road in the US. Transit is also hobbled by the space needs of traffic, as stops in a traffic-dependent landscape are surrounded by parking lots and rights of way for cars in traffic, not homes and workplaces for transit passengers. The same hierarchical road network that is easy to traverse for a car but difficult for walkers and bikers also makes transit routes torturous. Even if transit routes are straight along arterial roads, the hierarchical traffic network and spread-out traffic landscape makes it difficult for transit passengers to reach their transit stops.

Heavy and fast automobiles in traffic and trains or buses in transit are too dangerous if they collide, so steps are taken to prevent collisions at all costs. Roadways and rail transit rights of ways are almost always built in pairs of opposing ways, with clear delineation of which direction must travel on which side of the road. For the 4.5 million paved road miles in the US, there are 8.5 million lane miles, most built twelve feet wide. The only places where single-lane streets are built are places where traffic is one-way or traffic volume is low. Contrast this with sidewalks, where walkers are nimble enough to arrange their own lanes and collisions are low-impact and without incident. Bikeways are directional, like roads, but a two-way bikeway can fit in the lane width of a single traffic lane.

## The additional space cost of parking

Like any engineered work, roads are built for peak loads, not minimum, average, or median loads. It is a bad idea to have things we build fail under their heaviest expected load, after all. Roads and parking lots are built on that principle, sized for the peak hour of daily traffic, and the peak annual parking demand. For the other twenty-three hours of every day, and the other 364 days of any year, our roads and parking lots are bigger than necessary for the traffic that is on them. We have a paved area the size of West Virginia: getting hot, shedding rainwater, and making life hard for anyone without a car.

An essential part of any car journey is parking. You can't complete your journey and leave your car until your car is out of the flow of traffic, turned off, with the keys in your pocket. From your driveway to every place you work and shop, your car needs parking everywhere to be any use to you in traffic. Even though cars are marketed on drivability, roominess, or fuel efficiency, they spend most of their time sitting still, parked, with nobody in them. Zoning and subdivision regulations evolved as cities swelled with cars and traffic, mandating that all landowners provide sufficient parking for their land uses. Now, thanks to these parking regulations, 99 percent of traffic trips in the United States begin and end in a free parking space. Most paid parking is in historic downtowns built out before 1930. Streets were once surrounded by buildings and places to go. Now they are surrounded by parking and places to drive. New places charge for parking at their peril, as there are always other new places with free parking. A new development that charges for parking must be in high demand indeed.

Recently, my wife and I enjoyed some pizza at a restaurant in a suburban mall fifteen miles from downtown. You could tell the mall parking deck had visions of revenue, with gate arms and cashier booths, yet they have come to see that when 95 percent of the parking around there is free, they had darn well better also be free.

**I AM TYPING THIS IN A CAFÉ IN ONE OF THOSE HISTORIC DOWNTOWNS WHERE PARKING** is rare enough that you have to pay for it. I was unable to pay for parking on a broken meter in the cold sleet, and I didn't have a marker to deface the meter with the claim "BROKEN." If the meter reader takes her job as seriously as the mail carrier, I'll get a ticket. The parking ticket will be just a bit more than the cost of parking in a garage. I'll take the risk, as the parking space I found was much closer to my café than a garage, and probably free.

The space I parked in is not free to the city or its taxpayers. It costs them tens of thousands of dollars every year.[226] Parking spots are required by laws and calibrated to engineering standards. Animal clinics have different parking requirements than animal emergency rooms, swimming pool parking requirements are based on the cubic volume of the pool, and thousands of other locations are "parked" based on their particular form of land use.[227] All of these standards are documented by the Institute for Transportation Engineers (ITE).[228] While some of these categories are well-documented, others are not, which is a problem: either the developer or the party they sell the building to cannot recover the money wasted on extra real estate and pavement for unnecessary parking spaces.

Parking spaces are nine feet wide and eighteen feet long, big enough for cars, trucks, and SUVs (all close to six feet wide), and still leaving room for passengers to open their doors. The parking space needs to fit large SUVs and station wagons, so a Fiat 500 goes in the same 9x18-ft. (162 sq. ft.) stall as a Ford Excursion.[229] Cars need to get to and from their parking spaces, so parking lots are designed with ample circulation, and the width between rows is wide enough to accommodate one car getting out of a space and one car passing through. The average space per car in a parking lot is 325 sq. ft., or 0.8% of an acre, so 134 cars can be parked in an acre of parking lot.

The financial impact of parking requirements on urban lots is severe and can change high-density developments to medium- or even low-density, due to parking requirements that vary parking lot sizes based on the developer's stated use for the building. The different costs of surface, structured, and underground parking can force developers to build less leasable space than they initially planned, based on the price or rents expected from the building. The local market drives the rents that a building can command, but parking regulations are uniform across the nation. To bury parking under a building is orders of magnitude more expensive than clearing and coating a parking lot in asphalt. A developer cannot afford to build walkable urbanism with buildings right next to the sidewalk unless the location is likely to return the cost of building parking underground. If we want walkable places with pleasant views, without parking lots ruining the cityscape, every building must be built expensively and rented expensively, with parking under or within the building.[230]

Parking is a loss leader[231]—a private cost mandated by states and localities. For a surface parking lot, the developer must spend $1,000 to $2,000 to build each parking space. If the land is worth enough, and the owner can charge enough rent, the developer may build a parking garage starting at $10,000 a space, or underground parking at $40,000 a space.[232] Developers start considering parking garages when the cost of land is over $270,000 per acre. They start considering underground parking when land cost gets over $1.35 million per acre. These threshold prices are a function of the cost of land and the developer's ability to make money in sales or leases.[233]

One way to reconcile the public need for urbanism and the developer's need for a return on their investment is to uncouple or relax the parking requirement from the development type. This is anathema to many parts of America. Traffic has been the dominant transportation mode for about a century, connecting a sparse landscape spread out to the scale of traffic with parking, low-profile buildings, and stormwater ponds for the treatment of the inevitable runoff from all these roads, parking, and roofs. There is frequently anxiety about the feasibility of buildings without parking, but they usually outperform traffic-dependent places. The cost of accommodating parking and limiting access to automobiles only is greater than the cost of encouraging and building for foot and bike access. Enabling the construction of buildings without parking assumes that people can get to and from those buildings by means other than traffic, in the same way that parking requirements assume that people are required to get to buildings by traffic. Transportation and land use determine each other.

## Traffic and livability

The landscape we have built over the last century to accommodate traffic is mainly concerned with the needs of vehicles weighing over a ton and moving at an average of forty-five feet per second. There is no reason to think that a landscape built with that vehicle in mind should be good or suitable for people. Americans have been forced to hide from the landscape built for traffic if they want to find good quality of life and daily livability. Walkability has been relegated to subdivisions, shopping malls, and lifestyle centers.

AMERICAN CITIES DEVELOPED ZONING CODES AT THE VERY END OF THE 1800S TO separate industrial uses. They grew in prevalence and complexity at the same time as traffic did. They took on purposes like separation of residential from commercial land uses, which required residents to travel further to services they used to be able to find in their own neighborhoods. They also came to mandate separation of housing types, which implied class and racial segregation, either overtly or implicitly.[234] New neighborhoods were required to adhere to local zoning and separation of land uses, designed at a scale most usable in a car, at the scale of traffic.

Now, with a landscape built to the scale of traffic, parking is the most important amenity for every building. Houses are built with parking on the first floor, serving as the connection to the outside world. Retail and restaurants are built with a cartoon-like visual language meant to be recognized at a glance, the red-striped roof of the KFC, the green awning of the Starbucks, or the golden arches of the McDonald's. Nothing is meant to be walked to, only driven by. To walk or bike to anything has been made foolish in this landscape—far better to drive. If you cannot afford a car, the bus comes once an hour, as long as your job is near another bus stop. The only way to have choice in the landscape zoned for traffic is to have a car.

Traffic offers Americans the dream of sloth and detachment from surroundings beyond the roadway. Our dependence on traffic, and the engineering of all new places to serve it, has contributed to the epidemic of obesity in America.[235] We have engineered exercise out of America, and are the worse off for it.

Traffic is only usable by a segment of the population between young adulthood and old age. While this segment is most of the population, and the one that needs it the most to keep a job, this segment also makes all of us depend on traffic to get anywhere useful in the dawn and dusk of our lives. When elders' reactions and alertness become too unreliable, they become dangers to themselves and others in traffic when they get behind the wheel. My uncle had to drive my grandmother to her errands after she had a bad traffic collision. She no longer trusted herself to drive. She would have been stranded if not for my uncle, left completely dependent on paid drivers for her every need, until her money ran out. This stark choice is coming to tens of millions of American families with the aging population born after World War II. There will be sixty-one million Americans between the ages of sixty-six and eighty-four by 2030, most living in neighborhoods that can only reach services using traffic.[236] Traffic limits our choices to only traffic.

## Traffic and the built environment

By 1930 and the Great Depression, automotive traffic was widely viewed as a tool of prosperity and freedom. With Roosevelt's New Deal infrastructure programs, federal money became available to states and municipalities with concrete plans to use it for roads for traffic. Robert Moses, New York City parks commissioner, had a plan to use it, building roads away from and even through New York's historic neighborhoods, starting by 1936 with construction of the Triborough Toll Bridge. Adept at securing financing through local tolls and federal grants, Moses was a brilliant but frustrated progressive who worked to reduce urban overcrowding and allow people access to parklands. He became adept at New York state law and condemning land (even from the most well-connected landowners).[237] Moses's projects were opposed by many people whose neighborhoods were slated for right of way, like Jane Jacobs, author of *The Death and Life of American Cities*. Jacobs's opposition came about because of Moses's project to extend the Lower Manhattan Expressway through Greenwich Village and Washington Square Park. While working against Moses, she recorded her ideas about the social advantages of walkable and diverse cities like New York, and how they were ruined by freeways and traffic. Not until seventy years after Parks Commissioner Moses worked to remove people from the streets of New York did Streets Commissioner Janette Sadik-Khan work to invite people back to the streets of New York.

THROUGHOUT THE LAST CENTURY, TRAFFIC HAS HAD ITS SKEPTICS. ROADWAYS HAVE been built to manage traffic congestion and move people through neighborhoods, not create streets where neighbors can meet, converse, and play. That was the commonly agreed-on model of the street when everything in the street was moving at walking speed. The only refuge traffic offers for neighborhoods is zoned segregation and isolation from commercial services and jobs.

More recently, New Urbanism has challenged the design assumption of land use built out to serve only traffic. New Urbanist revisions to subdivision, parking, and zoning codes invite walking and biking back into American landscapes. Smart Growth principles have also enumerated the regional liabilities of dependence on traffic for all transportation. Both Smart Growth and New Urbanism suggest better ways to build transportation networks, as well as to improve the jobs to housing balance at the local and regional level, and to promote equitable housing and jobs policy, decreased reliance on impervious surface traffic lanes and parking, and human walk/bike-scaled zoning and development codes.

## Traffic costs to individuals

Transportation is the second-largest expense for most households, after shelter. Most cars last for less than a decade, so we buy several over our lifetimes.

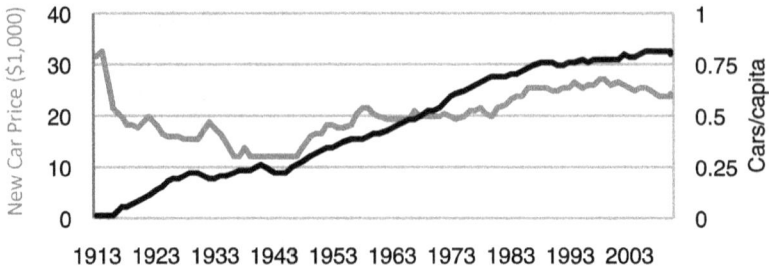

**Figure 4.6. Car ownership in America and inflation-adjusted purchase price from 1913 to 2010. Car ownership continues to expand to the present day.**[240] **The price of car ownership has risen sharply since 2020.**

The average price of a new car is over $25,000, usually paid with a three-, five-, or even seven-year loan with interest. Before the Great Recession, the average ownership term of a car or a house was around five or six years.[238] Now people are staying in their houses longer, waiting for home values to increase, but we still have to buy cars about as often as before, as they wear out mechanically and fashions and features change. A car is such a necessity to Americans to get around in the landscape we have built for ourselves that car ownership has increased, even with increasing prices (see figure 4.6). The supply chain issues caused by the COVID pandemic increased car prices greatly, but pandemic fears about urban overcrowding made car ownership even more appealing to American households despite the rise in costs, though the data is still forthcoming.[239]

Buying a car is the beginning of a long bleed of cash. The average cost of car ownership, estimated by the American Automobile Association (AAA), is 56 cents per mile. Multiply that by the twelve thousand miles the average car drives every year, and each car owner spends an average of about $8,000 on each car every year, as the admission price to traffic. Multiply that by the number of vehicles, and we spend as much as $1 trillion on operating vehicles in traffic every year in the United States. That is around 12 percent of the GDP (figure 4.7), behind healthcare's 18 percent and housing's 20 percent, but ahead of cheaper things like the military (5 percent), and environmental protection (2 percent).[241] Without depreciation, which is the cost of owning an asset that is losing value, the annual cost of car ownership for Americans is $850 billion. That is roughly equal to the amount added to the federal debt

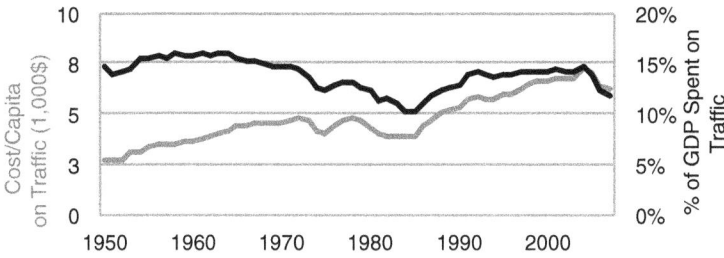

**Figure 4.7. Cost per capita of traffic (left axis) and percentage of GDP spent on traffic (right axis), showing a recent dip similar to the contraction after the OPEC oil crisis in the early 1970s set up by peak oil from America in 1971.**[242]

every year. The only way to spend less on traffic is to drive less, but for many Americans, that is not an option. Our jobs, housing, and shopping are too far away from each other, and our roads are built for us to drive more, not less.

Cars are large sunk costs. Once you buy a car, you would be wasting money not to drive it. The recurring costs of using a car eclipse the purchase price within a few years, but the initial shock is still there. If you spent several months' pay to buy the admission ticket to traffic, why not use it?

The COVID-19 pandemic disrupted many things about our local, national, and global lives starting in 2020. One thing it disrupted was global supply chains, both from the demand side and the supply side. In the summer of 2020, the price of oil and its distillates like gasoline fell sharply with the demand for fuel, as nations quarantined to "flatten the curve." As we developed vaccines and began to restart normal life, demand for goods like cars soared, while the pandemic was still killing millions around the world. Factories and whole cities were sporadically shuttered by quarantines imposed when transmission rates and hospitalizations got too high. By 2021, this combination of increased demand and sporadic supply started to cause general increases in inflation. Cars, and the gas required to fill them, have become a luxury good. Drivers can't switch to more fuel-efficient cars without running into supply chain delays and costs of new hybrid or electric vehicles.

Is this a reminder of the liabilities of traffic dependency, the way the horse epizootic of 1872 was a reminder of the liabilities of horse dependency?

The large sunk and operational costs of participating in traffic constitute a tax on every American living in a car-dependent place. This book will propose how to restore transportation choice to much of America. The fact that millions of American households are still paying off and using their cars in traffic makes this book's message hard to swallow for many, but it becomes more palatable when seen as a future of more choice. How many Americans are still paying off their cars from ten years ago? Hybrid, electric, and self-driving cars are all challenging the notion of carbon dependence and car ownership. The book also offers alternatives to car dependence by linking biking and walking to transit in the same way that parking is linked to roads.

## Traffic costs to governments

The gas tax once covered almost all the cost of road construction and maintenance. The feedback loop was perfect. The more people drove, the more they spent on gas, and the more they contributed to the construction of those roads, which would further incentivize people to drive more. And so the cycle should have gone on forever, a positive feedback loop for making infrastructure that served the needs of its users only.

There are two problems with this. The cost of maintenance has recently outstripped the revenues of the gas tax. Right now, the federal and state gas tax only covers 50 to 60 percent of the cost of highway maintenance (figure 4.8). The rest has to be covered from the general fund. The roads were supposed to be financed by a user fee, but are instead becoming an obligation of all taxpayers. Many states and localities spend as much on maintaining their roads as they do on schools, emergency responders, or police. Whereas those are expected expenses, the roads were supposed to be paid for through more clever schemes.

With rising gas prices, car owners have moved to curtail their consumption of gas, and therefore their payment of the gas tax. This is a great idea for household finances and the environment, but a problem for highways, transit, and bikeways funded through the gas tax. Some states are entertaining the notion of surcharging the owners of electric, hybrid, and small efficient cars like the Smart Car or the Fiat 500.[244] Because of the change in fleet fuel efficiency, reduction in vehicle miles traveled (VMT) since 2007, and legislative unwillingness to change the gas tax to keep pace with inflation, the US Highway Trust Fund is constantly on the brink of insolvency.

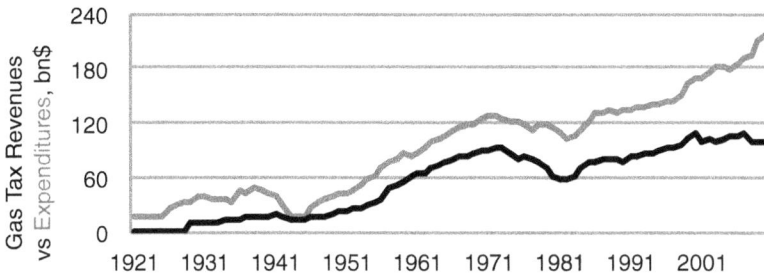

Figure 4.8. Gas tax revenues and expenditures for local, state, and federal highway budgets showing an increasing deficit between since 1995.[243]

A new idea in roadway financing is taxing VMT, not fuel. This is a great idea whose time has come, but its incentives are different from the gas tax. If VMT is taxed directly, without accounting for vehicle size, then larger vehicles, which put more wear and tear on the roads, could get a free ride. The VMT tax only works if it accounts for vehicle mass, however. A proper weight-adjusted VMT tax could account for everything down to bicycles, which might pay a penny or two every ten miles while larger trucks could pay a dollar or more per mile driven. The VMT tax would punish long-distance commuters and travelers, who currently use less gas per mile for long-distance, high-speed driving than shorter-distance travelers in congested stop-and-go traffic.

The simplest way to measure a single vehicle's miles traveled is by comparing the mileage at registration form one year to the next. This does not account for where the vehicle was driven or if it went out of state to wear out road built by other states. There are privacy concerns with using transponders linked to specific cars with specific drivers at specific addresses, but they would be able to distinguish between in-state and out-of-state driving. Americans have been enthusiastically adopting transponders for the last fifteen years, in the form of smart phones with GPS location services. Transportation data analytics like StreetLight collect smartphone movements and infer transportation mode and purpose-based speed and time of day with astonishing accuracy.[245] It would not be difficult for state DOTs to ask citizens to install an app on their phones to collect data on their movements on state and local roads, sidewalks and bikeways to bill them for wear and tear on transportation infrastructure.

Local communities pay for local services like schools, fire, and police with sales and property taxes because these things improve their local value. They are naturally tied to their communities. The better your local services, the better your property value. The more tax revenue your local government collects, the more they provide better services like libraries, museums, and sports. Roads, while they serve the needs of locals, are just as likely to allow outsiders to move through without stopping. Roads are more properly the jurisdiction of state or regional governments. Nevertheless, road maintenance for traffic is a significant expense for local and state governments. Road maintenance is only getting more expensive as roads age past the last great pulse of road-building in the 1960s and 1970s.[246]

## Conclusion

*The Ways of the World* by the traffic engineer Maxwell G. Lay is a history of all things related to roads. In it he postulated a law for the behavior of traffic: "If you have a car, you will use the car as often, as much, and as far as you can." If that law is true, what is the point of this book? How do we overcome that? If traffic is the highest and best use of transportation, what use are transit, biking, and walking?

The following chapters compare traffic with transit, walking, and biking to understand why it is possible and important to overcome Lay's law in carefully chosen places.

# 5

# How Travel Is Changing

This chapter is an exploration of the current state of transportation in the United States, setting the stage for a recommendation of a better way. I am not interested in excoriating or ignoring traffic. It is too powerful, ubiquitous, and useful for that. I am interested in outcompeting traffic. To get to something better than traffic, we must see where other modes are already better and how that might affect what we do in the future.

If traffic is both an inevitable boon of American life and a terrible scourge, what do we do? Traffic's dominance is not a conspiracy against the sustainable alternatives of transit, biking, and walking. Rather, we have built to accommodate it over the last century to make it more compelling. The new geography of American metropolises, the avarice of trolley barons, and the obvious need for motorbikes to any hill climber of the 1890s meant that America would embrace traffic over transit earlier than any European nation. While Europe explored traffic in the 1950s and 1960s to escape the bicycle and transit, America used traffic in the 1900s and 1910s to escape the filth and danger of the horse and never looked back.

My aim is to show how we could restore choice between traffic, transit, walking, and biking. In this chapter, I discuss public transit in general, because data for that is the most comprehensive. Later in the book, I profile rail transit in particular, even though it accounts for fewer than half of the transit trips in the United States today. I acknowledge that there are fewer than fifty cities with any form of rail transit. Most rail transit (commuter, heavy, light, monorail, automated, and funicular) has distinct routes, separate from traffic, and performs much differently than traffic. Bus transit provides over half of transit rides in the United States, but the performance of bus transit is not as distinct from traffic as rail. For the sake of simplicity, I'm only considering cities with heavy, light, or commuter rail lines and stations, not streetcars, people-movers, or monorails.

The challenge is proposing something that lets all four transportation modes work well at their own scales. A walkable, bikeable place, like the downtowns of cities that were large and established in 1900, is not a great place to use traffic. Most of the rest of America is not a great place to walk or bike, because of traffic and its implicit threat of collision, injury, and death. Too many buses gum up traffic worse than delivery vans with their constant stopping and starting to pick up and deposit passengers. For much of the twentieth century, the solution and program of infrastructure was to build everything for traffic. We may be entering a new period of skepticism of traffic, but we need to be explicit about what we want done differently and what makes sense.

## Trends in transportation at the national level

One of the trends in recent years has been a decline in car vehicle miles traveled, down to 1989 levels. Since the beginning of the Great Recession in 2007, Americans have simply been driving less (see figures 5.1 and 5.2). They are also driving more alone, according to data on vehicle miles traveled (VMT) versus personal miles traveled (PMT).

Per person VMT in 2011 were at the same level as they were in 1997, and the per person PMT in traffic were the same as they were in 1991. VMT per capita have never fallen this much in the US.

What's going on behind this decline in the number of miles traveled in traffic? Are we using transit, walking, and biking more? The National Household Transportation Survey (NHTS) lets us see in detail what's happened. NHTS has surveyed household transportation choices every decade or so since 1969, and began to track biking and walking in 1995. The great thing about the NHTS is that it surveys all trips and sorts them by length, mode (e.g. walking, biking, traffic, transit, etc.), purpose, and other factors. We'll look at this data to pinpoint what has been happening to America's transportation, and the transportation modes and the ages of travelers that have changed the most. We need this because when devising any new approach to transportation,

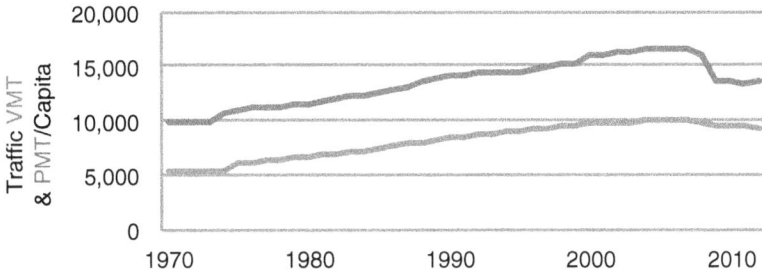

Figure 5.1. Vehicle miles traveled (VMT, gray) and passenger miles traveled (PMT, light gray) per person, declining since 2007.

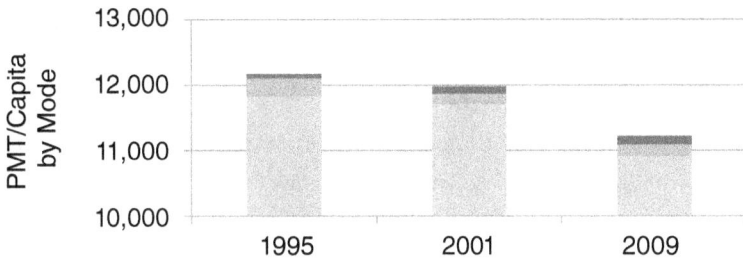

Figure 5.2. Passenger miles per person by mode, 1995–2009. Light gray is traffic, dark gray is transit, and the dark shade at the top of each bar is walking. Note that the scale starts at 10,000 PMT to account for the large amount of travel by traffic.

we're much more likely to be successful if we go with the flow of how people are already changing their transport choices.

THE IMPORTANT UNIT OF TRANSPORTATION IS NOT THE MILE TRAVELED, BUT THE TRIP, because a trip represents someone going out to get what they want: A trip implies purpose and completion of that purpose. If what they want is a walkable distance away, and they walk one mile, they are just as happy as they would be if they had driven ten miles in traffic for the same purpose.

The data in figure 5.3 show that between 1995 and 2009, walking trips per person increased by 96 percent, and biking trips per person by 48 percent. Though traffic trips still represent the lion's share of all trips taken, traffic trips per person actually declined by 14 percent during the same interval. Reliance on transit meanwhile declined by 8 percent.

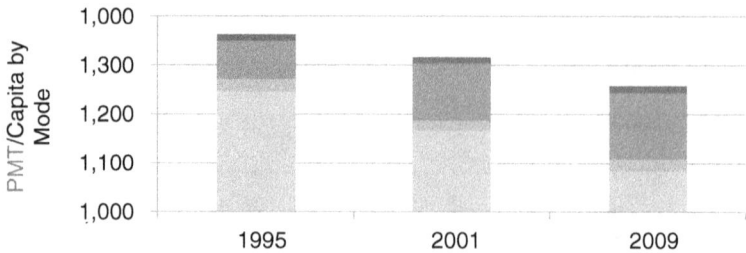

**Figure 5.3. Trips per person by mode of transportation, 1995–2009.**[247] **As in the graph above, traffic is the bottom stack, transit the middle stack, and walking the next stack with the top stack showing biking.**

To better understand these trends, I examined the data at a finer level of detail by the distance of trips and by age group making the trip. Distance of trips is valuable in understanding mode choice because not all modes are suited to all trips; the average trip distance is ten miles, feasible only by traffic or transit. But what about the trips that could be traversed by walkers or bikers, if the conditions were right? The average bike trip is three miles. Although bike trips can be as long as twenty or even one hundred miles, only the most dedicated cyclists make these trips, and seldom for daily travel. Similarly, walking trips are frequently under a quarter mile.

Therefore, let's examine changes in the number of trips per person from 1995 to 2009 for trips under nine miles (transit), three miles (bike), and one mile (walk), and we'll look at the change by age group in order to understand how many trips could be completed by non-traffic modes, by what age groups, and how this has progressed. Even though these trip distances do not say anything about mode, their distances indicate if they could be reasonably be made by non-traffic modes. Figure 5.4 compares trips under nine miles traveled per person by age group for each NHTS year.

The pattern of trips under nine miles is much the same as for all trips, though they represent about half the trips. What about the trips that would be most suitable for a bike, approximately of fewer than three miles?

Figure 5.5 shows a steady decline of trips of less than three miles between 1995 and 2009. This segment represents about a quarter of all trips. What about trips that are within reasonable walking distance, less than one mile?

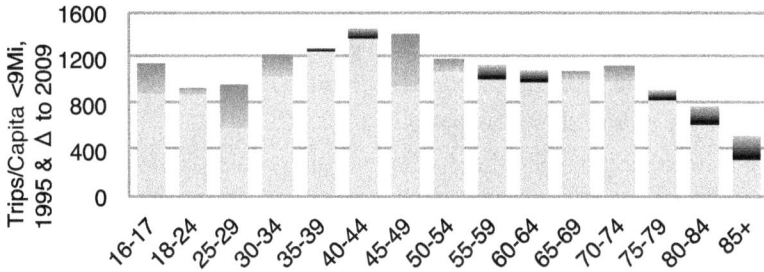

Figure 5.4. Change between 1995 and 2009 in all trips fewer than nine miles per person by age group. Darker tops of columns show an increase from 1995 to 2009, while darker middles of columns show a decrease.

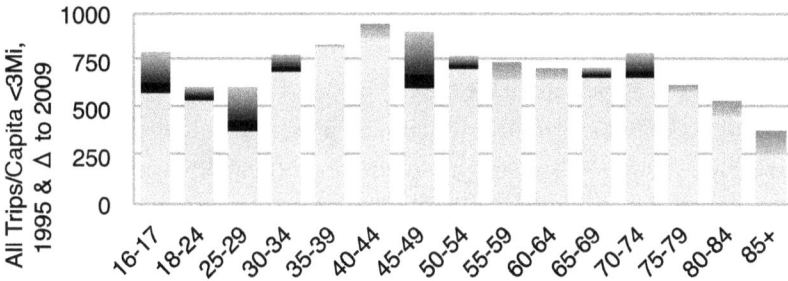

Figure 5.5. Change between 1995 and 2009 in all trips of fewer than three miles per person by age group. Darker tops of columns show an increase from 1995 to 2009, while darker middles of columns show a decrease.

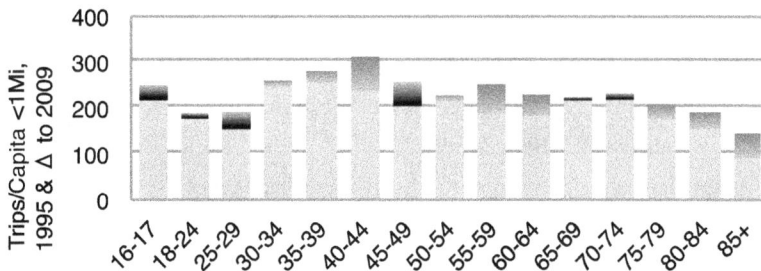

Figure 5.6. Change between 1995 and 2009 in all trips less than one mile per person by age group. Darker tops of columns show an increase from 1995 to 2009, while darker middles of columns show a decrease.

Trips under a mile that people could easily walk represent almost 17 percent of all trips. Interestingly, we see an uptick in these trips, particularly in middle age (figure 5.6). Trips increased considerably for 35–45 year-olds and 55–65-year-olds in 2009 from prior years.

While the graphs above help us understand relative trends in per-person trips by age group for distances in which biking and walking could potentially compete with traffic as a viable transportation mode, they do not distinguish which mode people are using within those defined distances. The next section drills deeper to understand trip trends by mode. Using the same data as above, what do we see for modes best suited to trips under nine miles, three miles or one mile, like transit, biking, or walking? Most importantly, we can also show trends between 1995 and 2009 in mode usage, using a metric that translates across modes.

Traffic is the most prevalent transportation mode in America, accounting for about 80 percent of the trips in the US. Looking at the age distribution we see a decline in traffic trips among almost all age groups between 1995 and 2009 (figure 5.7). The only groups to increase their driving were the baby boomers and older groups.

The next mode, transit, is the only alternative many consider to traffic. For transit, total trips show a rebound from a 2001 low, but remain lower than transit use in 1995. Transit use in 2009 is lower for most ages than in 1995, with some increase for ages sixteen to twenty-four and thirty to thirty-five (figure 5.8).

This is consistent with a national decline in transit use since World War II. It is also a national trend, abstracted from the specifics of certain metros, cities, or even routes. The changes between 1995 and 2009 look almost like the opposite of traffic, but the magnitude of these changes is about 1 percent of the magnitude of the changes in traffic.

At similar volumes but more uplifting are the changes in biking between 1995 and 2009. Biking is rising dramatically but represents less than 1 percent of the total trips in the United States. Again, note that like transit, the volumes are much smaller than traffic. Unlike transit, no age group significantly abandoned biking, and the middle age group took it up a great deal more in 2009 than in 1995 (figure 5.9).

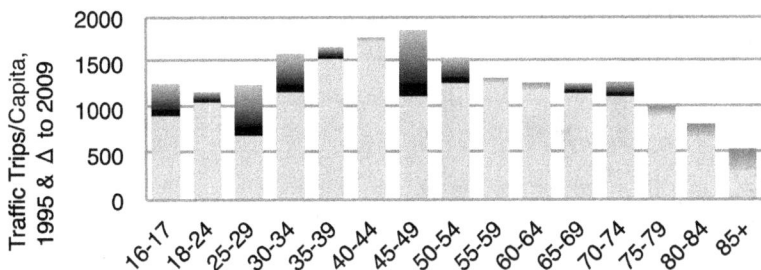

Figure 5.7. Change between 1995 and 2009 in traffic trips per person by age group. Darker tops of columns show an increase from 1995 to 2009, while darker middles of columns show a decrease. Note that traffic trips decreased for most age groups in this period.

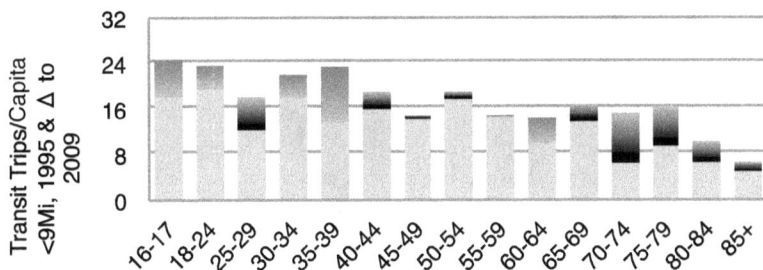

Figure 5.8. Change between 1995 and 2009 in transit trips fewer than nine miles per person by age group. Darker tops of columns show an increase from 1995 to 2009, while darker middles of columns show a decrease.

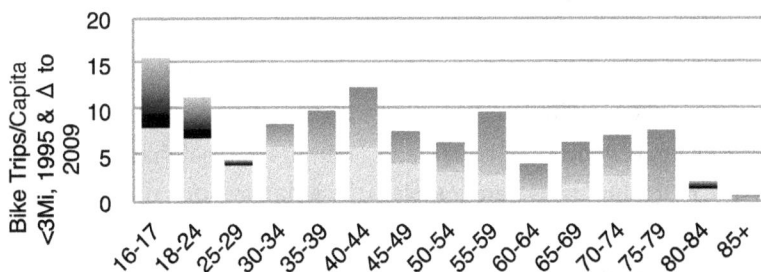

Figure 5.9. Change between 1995 and 2009 in bike trips <3 miles per person by age group. Darker tops of columns show an increase from 1995 to 2009, while darker middles of columns show a decrease.

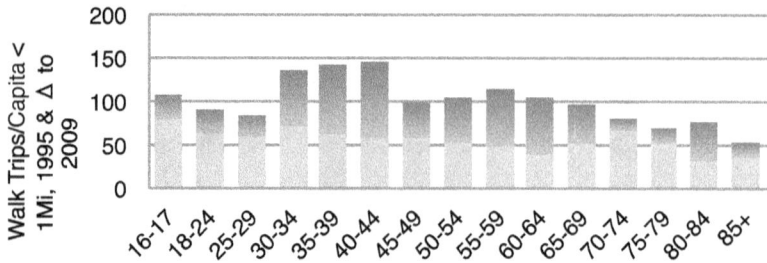

Figure 5.10. Change between 1995 and 2009 in walking trips <1 mile per person by age group. Note that every age group increased per capita walking trips, more than doubling walking trips for some age groups.

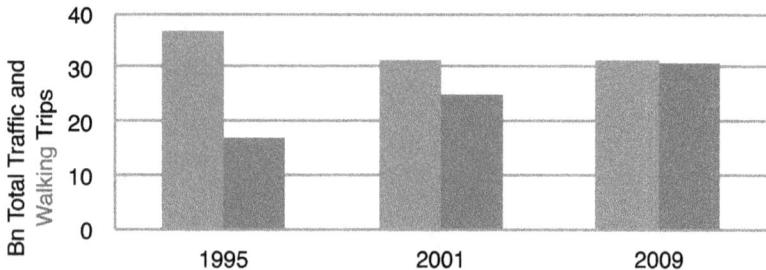

Figure 5.11. Comparison of traffic and walking trips of less than one mile per person in 1995, 2001, and 2009.

Biking has also declined for the young in this survey period, possibly off-set by greater use of traffic at that age.

Finally, walking trips account for 15 to 20 percent of all trips, and grew for all ages between 1995 and 2009 (figure 5.10).

In fact, the increase in walking over the fourteen years is so large, I wanted to compare the use of walking as a transportation mode versus traffic for trips of a walkable distance: that is, less than a mile. It turns out that for many age groups, more short trips were completed by walking than traffic (figure 5.11).

This is one of the most important findings in this chapter. It is not only a first for the period of the survey, but perhaps also a first for the last few de-cades of traffic dependence. Looking at the data a different way, about half of the trips under a mile that Americans take, they now take on foot. This is an interesting result, and one I will build upon in chapter 6 and beyond.

There have been nationwide revolutions in walking, biking, and traffic since 1995. Americans are making fewer trips since 1995, an artifact of better trip chaining and the rise of the internet as a force of convenience and connection. Traffic trips and miles traveled have declined for the second time in history, and much more than in 1973. Walking has increased immensely, especially among the middle-aged. Biking and transit, each accounting for around 1 percent of all trips, have increased also, but they remain minority players, nationally.

Short trips make up a surprisingly large portion of total trips in America. We take most of those trips in traffic because the infrastructure is most built out for traffic. Where Americans could take short trips on foot, bike, or transit, the way is often blocked or inconvenient for the sake of traffic. The changes between 1995 and 2009 show that we are willing to take more short trips without traffic. How and where we do this is useful to consider for existing conditions and future proposals.

## Trends in transportation at a local level

National trends do not tell us much about what is happening where you or I live and work. Place is important when it comes to transportation choices. Forty percent of all the transit trips in the United States are in New York City, leaving much of the rest of the nation wondering what transit is for. What does transportation look like at the local level and to what extent does place matter in influencing mode choice? The census has been surveying the commute since 1960, and surveying walking and biking as modes since 1990. For this book, I surveyed the commute data from the 2010 Census.[248] This provides a conservative estimate of where people are using different kinds of transportation.

Information about the journey to work, while limited, is sufficient to understand where Americans like to bike, walk, or use transit. If a place has a lot of people biking or walking to work, there are workplaces that it is safe to walk and bike to within walking and biking distance of their homes. A lot of America cannot make that claim.

I identified the block groups that took over 10 percent and over 50 percent of the walking, bike, and transit journeys to work. A place isn't *really* bikeable unless over half of the trips are by bike, but even 10 percent of trips is still significant.[249]

The two major problems with "journey to work" as a measure of transportation is that we might not use just one mode of transportation. We might walk a long way from parking to our job; park and ride and take a bus on highways for the rest of the journey; walk to transit and take a bikeshare bike from the station to work; or drive in the morning and carpool back in the evening on alternate days. The notion of one mode of journey to work supposes that we can move directly from our homes to our jobs, a convenience most enjoyed by walkers, bikers, and traffic. The other problem with evaluating journey to work is that it is only about 20 percent of the travel that we do, by miles. This is somewhat unfair, as 30 percent of the miles we do travel are on irregular long trips over fifty miles in length. However, by trip, not miles, the journey to work is more important.

The journey to work is the most important, and best to measure, because it is made when traffic is the most congested. It is when most traffic users feel the pain of traffic. It is also the volume that highway engineers design roads to handle. Roads are designed to handle peak traffic for fifteen minutes in the morning and afternoon, and are overbuilt for the rest of the day, inviting traffic to move faster. This makes things dangerous for bikers and walkers who have to move along or across those roads. To be fair, sidewalks, bike trails, and transit stations are also sized for their peak demand. The sizing, and subsequent oversizing, of roads so that they can accommodate peak traffic is the most pervasive and important infrastructure choice we make. The following maps show block groups where transit, biking, and walking were used for over 10 percent and 50 percent of work trips in the 2010 census.

Not surprisingly, transit gets more use the closer we get to cities (figure 5.12), particularly densely populated metropolises such as San Francisco, Los Angeles, and Seattle on the west coast, and Boston, New York, Philadelphia, and Miami on the east coast. Block groups with over 10 percent of transit share for work trips comprise 15 percent of all block groups but represent only 1 percent of land area in the US.

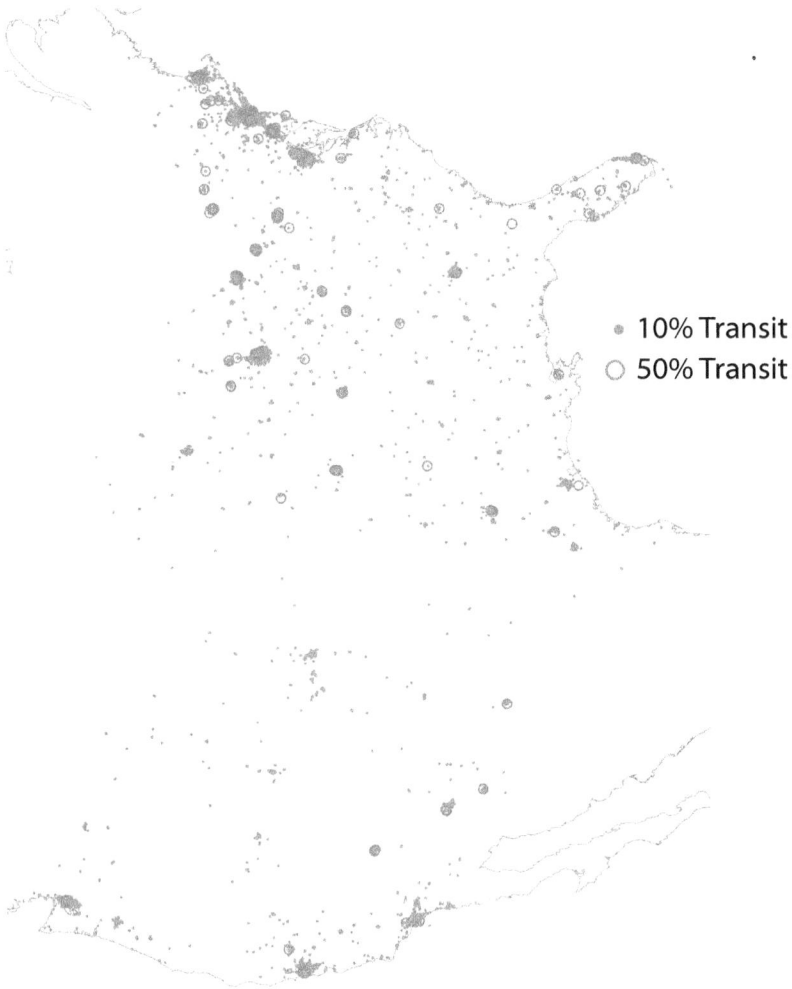

Figure 5.12. **Block groups where transit represents more than 10 percent (dot) and 50 percent (circle) of trips to work (2020).**

Biking is even more concentrated. Only 1 percent of all block groups in the US show more than 10 percent of commutes by biking. There are only seventeen block groups in the entire country where biking is used for more than 50 percent of trips to work (figure 5.13). Proximity to universities, including UC Davis, UC Santa Cruz, and Stanford, appears to be a force for a majority of

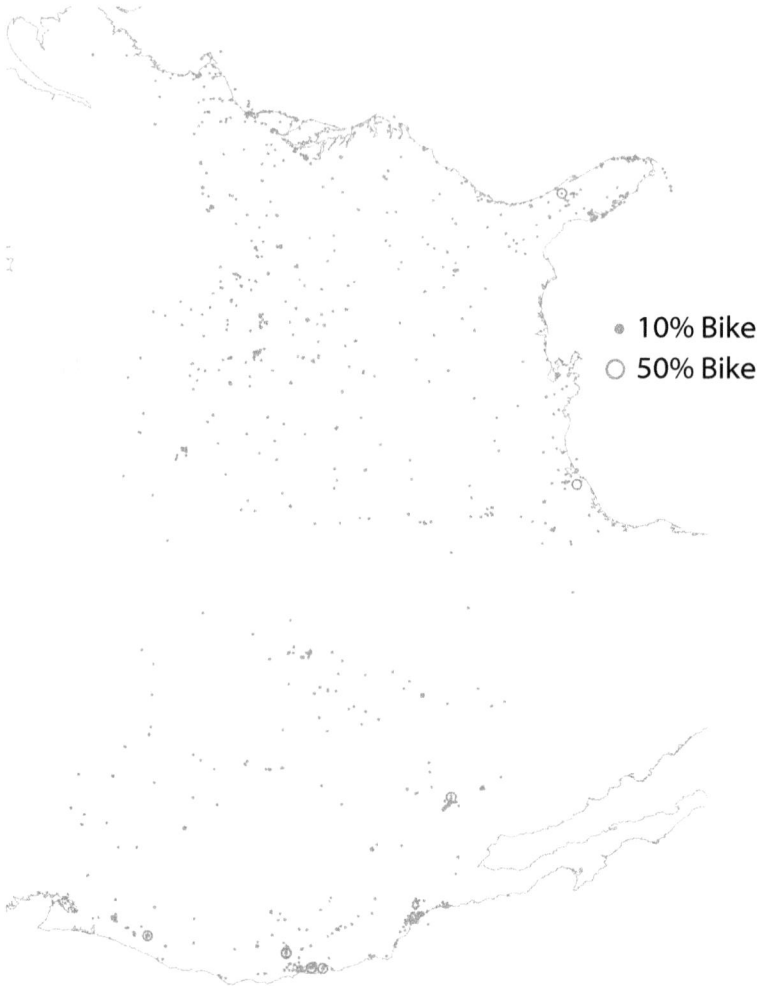

Figure 5.13. **Block groups where biking represents more than 10 percent (dot) and 50 percent (circle) of trips to work (2020).**

these biking block groups. Otherwise, biking's dominance is limited to cities such as Portland, New York, and Denver. However, even the most bikeable city in America, Portland, reports less than 10 percent of its workers biking to work.

In contrast, the incidence of block groups where Americans use walking as their mode for journeying to work is more widespread and prevalent than expected. A full 8 percent of all block groups in the US have over 10 percent

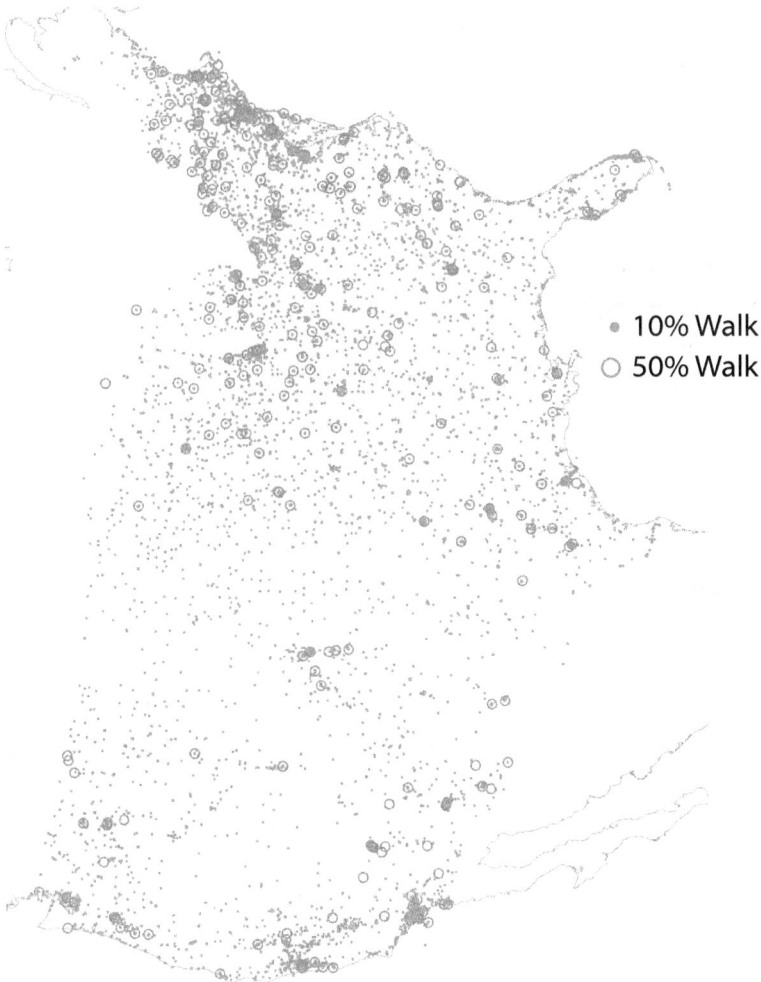

10% Walk
50% Walk

**Figure 5.14. Block groups where walking represents more than 10 percent (dot) and 50 percent (circle) of trips to work (2020).**

of walking trips to work. Though the incidence of walking grows the closer we get to cities, walking trips are also used well outside of cities. Unlike transit or biking, there is no state in which there is not at least one block group where walking is used for more than 10 percent of work trips.

As we zoom in from state to metro to city to census tract or block group scales with the journey to work statistic, we find thousands of places where walking, biking, or transit are over 10 percent or even 50 percent of work trips. As much as we think traffic is the only way to get around America, there are places throughout America where traffic is a choice, not a mandate.

## Places where traffic isn't dominant

Looking at the land uses in areas where traffic is not the absolute default does provide some insight, however. Together these block groups represent 11 percent of the total land area in the US and 15 percent of the population. I also sought to understand how much of America's economy happened in these places by looking at the housing and jobs in each block group. I wanted to see if I could understand what Americans are biking and walking *to*. How many jobs and housing units are in the kinds of places where walking, biking, or transit use make sense for a significant share of commuters? The same 11 percent land area represents 15 percent of housing units in the US and 20 percent of total jobs.

To further examine this, I sorted the block groups into those within a mile of a college or downtown,[250] and those further away. It turns out that at least 70 percent of block groups where the share of walking and biking is over 10 percent are within a mile of a downtown or university. For the walking share alone, more than three-quarters of these block groups are within a mile of a downtown or university. These walking-share block groups contain about 93 percent of the jobs of all block groups with greater than 10 percent of the walking share in the nation. For block groups with greater than 10 percent of the bike share, the percentage of jobs within a mile of a university or downtown is closer to 86 percent. Grouping this data by states, block groups with a higher incidence of walking and biking trips to work are predominantly located in fifteen states (figure 5.15).

Some states, like New York or California, have a lot of walkable, bikeable, or transit-accessible places; they are also large states. The populations of these places represent a small fraction of the state population, most of whom use traffic to get to work.

Intensity is the sum of housing units (HU) and jobs per acre in a block group. This is a better metric than population density. Intensity lets us look at the overall economy of a place in terms of both incomes and homes. For the

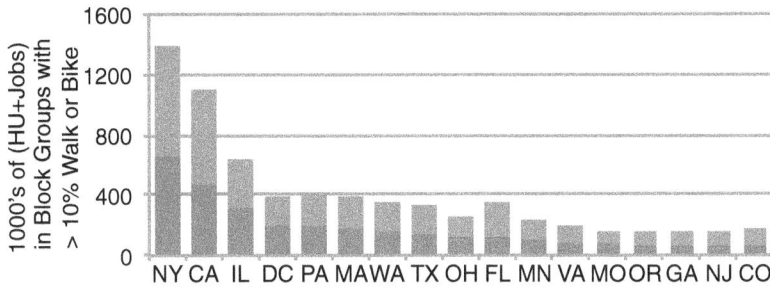

**Figure 5.15.** States with 80 percent of the nation's block groups with over 10 percent of the share of walking or biking to work. About 40 percent of these block groups are within a mile of a university or downtown.

whole United States, there were 242.3 million Housing Units+Jobs in 2010, for an overall intensity of 0.11 (HU+Jobs)/acre. For block groups where the transit to work trip share is over 10 percent, intensity increases thirteen times to 1.5 (HU+Jobs) per acre. For block groups where the biking or walking to work trip share is over 10 percent, intensity increases further to over 14 (HU+Jobs)/ acre, over one hundred times the national average intensity. Many of the most walkable and bikeable parts of America are near historic downtowns or university campuses. The national average intensity includes open land and forest. Within a mile of these nuclei, there are 27.2 million HU+Jobs in block groups where the share of walking or biking share is over 10 percent. This 11 percent of America's households and jobs is in places where biking and walking are a common way to commute.

Land use choice plays a definitive role in determining transportation choice. That is, in order for transportation choice to make sense for a particular trip, people require a certain scale that is workable for that mode of transportation—somewhere between one to three miles for walking and biking—and some level of density of development that allows them to fulfill the purpose of the trip. The data supports economic land use choices foremost, such as residential and business where people live and work. Due to the affinity of walkers and bikers for downtowns and universities, however, we can also surmise that related land use choices are applicable. People living and working in these block groups may also wish to include recreation and retail land use, which enable secondary economic activities such as errands, dining, and shopping.

These thousands of American places where walking, biking, or transit are reasonable choices for commuting indicate how we might pull transit and biking off the floor of usability to let them complement and compete with traffic. The remainder of this book shows how we might do it, and what we gain.

# 6

# The Solution

The point of transportation is getting where we want to be, on time. Nothing in that definition says we need to drive there. Our standards of how far can be traveled and who can travel have changed over the centuries with technology and economy. What if "there" was an easy walk from "here" for more of America? What if every transit stop was surrounded by enough businesses and homes to make the stops an easy walk for people living or working nearby? What if people could walk and bike where they were going near transit stations more easily than driving? What if transit station areas were built up enough to be both origins and destinations?

## Proximity as transportation

The answer is proximity.

The compelling reasons to ride transit are that it stops within walking distance of your destination and that you can walk to a stop from home. You don't need to get into your car at all. Instead of wandering through acres of parking lots, you step out of the transit station onto streets of businesses, homes, and parks, all within walking or biking distance. More doors closer to transit stations and to each other make every rail and bus station and stop a focus of jobs, housing, and shopping. No matter how frequent or luxurious transit service is, if it doesn't connect where people are to where they are going, many won't use it.

With proximity, there are things worth walking to. Where there are things worth walking to, there is little need for a car. Where there is proximity there is less need for a massive traffic system. Proximity of housing, jobs, recreation, and services to every station on America's regional transit networks could revive transit, biking, walking, and real estate markets in the US. Traffic is great for long-distance, family, or hauling trips. Over the last century, we have used traffic for trips better suited to transit, walking, or biking. Trips under three miles could easily be done on a bike, if they were safe from traffic. Trips under a quarter mile or even a mile could easily be walked, if the walk was worth taking and not on the side of a road or highway. The solution proposed here gives America the best of both worlds—traffic for where it's really needed, transit/walking/biking in places where they work better.

To enable walking and biking in America, we need laws, regulations, and standards that allow places for walking and biking to be safe from traffic. Just as traffic-dependent places offer parking to weary drivers, walkable places need to offer storefronts, housing, and things for walkers to do and see right up next to the sidewalk. Bikeable places offer lanes and paths that are distinct from trafficways and walkways, with plenty of things within biking distance. Zoning requires parking on private property for any land use connected to the road and traffic engineering requires a new lane of roadway for every eight hundred cars/hour. We should require more capacity for bikers and walkers within and around places suitable for walking and biking. Just as traffic engineers assume they will get the full use of through lanes, with parking handled by every property owner adjacent to every road, walking/biking engineers should assume that things will be built with enough intensity and choice of land use to enable people to walk and bike in comfort and safety. This has been easy in places built a century ago but prohibited by zoning standards in nearly every place built since.

Let transit, walking, and biking succeed in America by having them affect land use and transportation circulation that same way that traffic does. Just as traffic needs wide lanes, curb cuts, and ample parking on private property in order to function, walking and biking need things to do within walking and biking distances and safe paths between them in order to function. Transit needs to connect many people at one station to many places at other stations, and functions best when those things are within walking and biking distance of every station.

**THE CONVENTIONAL AREA OF INFLUENCE, UPZONING, AND DEVELOPMENT AROUND** transit stations has been the distance an average American walker can walk in fifteen minutes and five hundred steps: a quarter mile. This is not enough. The capacity of transit for tens of thousands of passengers per hour supports more development than can be squeezed into that quarter-of-a-mile radius (a fifth of a square mile). Beyond that quarter mile could be walkable urbanism or the conventional American landscape of highways and parking.

The average American walker can walk a half mile through an IKEA to the cafeteria, because there are things to see and do along the way. With the capacity of transit, the area around transit stations could be a walkable origin and destination for residents, workers, shoppers, and people within a mile of the station or more.

**THE ONE-MILE WALKING CIRCLE AROUND EVERY TRANSIT STATION IS AN IMPROVEMENT** over the quarter mile, but it only works with enough density and diversity of housing and jobs, origins and destinations, to make that walk worthwhile. For a place to be walkable, jobs that pay enough to afford what housing costs should be within walking distance of each other. For a place to be walkable and bikeable, it needs pathways within and between places, usable only by bikers and walkers, and safe from being killed by traffic. This is the fifteen-minute place, where nearly everything we need is within a fifteen-minute walk of where we are.

While considering the fifteen-minute place, why not consider biking as well as walking? What if we enabled places within a fifteen- to twenty-minute bike trip of transit stations—about three miles—to have just as much development intensity and diversity as walkable places? This three-mile radius, twenty-eight square miles, would be linked to the rest of the station area by bikeways and shared use paths, just as the area closer to the station would be linked by bikeways and walkways.

Allow, enable, and build fifteen-minute places at both the walk and bike scale around every transit station. Americans would then be able to get almost everywhere they need within their home station area or a different one within their region.

**TRANSIT CAN OFFER HOUSING AND TRANSPORTATION CHOICE, BUT ONLY IF WE TAKE** land use around transit seriously. We need to allow bikeable and walkable routes and land use intensities within a bikeable three-mile trip of all stations, not just the quarter mile. If sparsely located transit stations aren't providing connections to the last mile of most trips in your region, bring the last mile to the transit stations. Transit should allow passengers to walk out of stations to home, work, shopping, or play, without the daily need for traffic. Within those three miles, zone and build for more diversity, greater intensity, and less traffic. Rebuild existing traffic arterials to radially serve the transit station or avoid the station area entirely. Traffic arterials should allow space for all modes to feel safe from traffic. Build new connections independent of traffic to enable walkers and bikers to move safely within the intensified station area. Transit station areas should be built to serve walkers, bikers, buses, and traffic, not only traffic and buses. America's laws have forced traffic to influence land use for nearly a century. Now, it is time to consider how to let transit, walking, and biking influence land use as well. Just as every car deserves five to eight parking spaces, every transit station deserves twenty-eight square miles (a circle with a radius of three miles) of bikeable urbanism around it.

## Where can we do it?

Proximity does not need to be the default policy for all the land, only the bits within a fifteen-minute bike trip of transit. This proposal does not compel land use beyond those circles of influence, in the same way that parking regulations do not influence land uses not connected to the road network. Let the car-dependent landscape stay car-dependent. Zoning laws since 1899 have dealt with the spacing of the landscape to the scale of traffic. Zoning has used traffic's range and speed to separate land uses and people by distances only usable by traffic, not walkers or bikers. It is time to let all transportation modes affect land use, and to restore choice to the American landscape. Housing choice should not be a quaint historic accident, but nationally available. Give every transit stop the type of development it needs, and we revolutionize the American real estate market, while making transportation more affordable and safer. Where there is transit, let it support as much housing and as many jobs as it can, based on the number of seats offered by the transit route.

Only twenty-nine American metropoles have light, heavy, or commuter transit lines or networks. This book proposes changing default zoning around the 4,096 stations of these transit systems. Allowing transit to affect land use as much as traffic could also be applied to the hundreds of thousands of stops served by eight hundred bus transit networks in America. Bus transit does not have the same capacity as heavy rail transit, but the areas around its stops should still be focused on making those stops walkable, bikeable origins and destinations. Land use should be a forethought in transit stop and station development. Land use is already a forethought for traffic network design, as described above.

A little over half of Americans live and work in the metropoles with rail transit systems. That share is growing, as these are the most economically dynamic and diverse metropoles in America. These are also the metropoles with the worst issues of congestion, housing affordability, and jobs to housing balance. Our proposed solution offers cities that have installed systems of rail transit ways to restore choice to their transportation, housing, and jobs markets.

The area within twenty minutes' biking distance of all rail transit stations is only 1 percent of all the land in the US. Only 0.03 percent of the United States is built to the walkable intensity of fourteen housing units plus jobs per acre (Intensity Units [IU]) (see chapter 5). Most of the areas built out to that walkable intensity were built before zoning for traffic made it illegal. While most of the transit station areas in historic city centers were already developed at walkable intensity or above, many suburban or newer stations were not (figure 6.1).

What if we upzoned and redeveloped for that walkable intensity of 14 IU in the 1 percent of the US within biking distance of a rail transit station? Figure 6.2 shows how much potential development we could have on the land around existing transit stations if we allowed them to be built at walking and biking intensities.

The potential for development around stations is striking. If we built up the three miles around all our transit stations, we could add 50 percent more jobs and housing to the US. Walkable intensity is not dense urbanism. It is

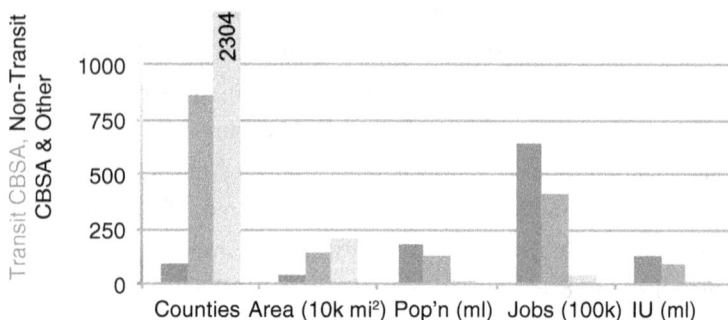

Figure 6.1. Area, population, jobs, and intensity of Core-Based Statistical Areas (CBSAs) with and without transit, compared with non-CBSA.[251] CBSA with rail transit are in gray, CBSA without rail transit are in dark gray, and Non-CBSA are in light gray. An Intensity Unit (IU) is one housing unit or job per acre.

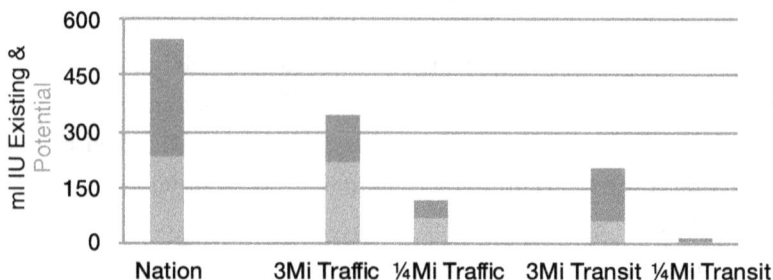

Figure 6.2. Existing and potential IU in US, near traffic, and near transit.[252] Gray areas in columns show potential for development to 14 (HU+Jobs)/acre.

comparable to the inner ring suburbs built in most American cities between 1890 and 1930. This is the intensity we forgot to build for when we decided to build for traffic and not for people. We gave traffic control of land use a century ago. What can we do if we allow transit to do the same?

## Making the change

What does a walkable and bikeable station area mean? Right now, most station areas are built only for traffic, with parking and curb cuts dominating the landscape. The rest of this chapter suggests possible steps to transform a station area from traffic-dominated to walkable and bikeable. The example shown is the Vienna/Fairfax metro at the western terminus of the heavy rail Orange line in the Washington, DC, metro.

First, a map of the current zoning shows a landscape made to be used by traffic (figure 6.3). Walking from residential to commercial areas is inconvenient by design.

In this existing zoning, there are already some areas of walkable intensity, shown in the shaded block groups. The roadways and buildings in this suburban area are not built for walkability, but traffic.

Figure 6.5 shows the roads and green spaces around this transit station.

## Step 1: Implement more "connector routes" for bikers and walkers

The first step is diversifying the local route network for bikers and walkers. Enhance the existing road network to make the landscape more usable for walkers and bikers, making a wide variety of destinations safely available to walking and biking, not just traffic (figure 6.6). The bike and walking network of pathways is woven together within and between the existing traffic networks. Instead of twelve-foot lanes, walking and bike paths only need four- or six-foot lanes. Where walking or bike lanes are next to traffic, they should either be fully separated or the speed limit of traffic should be reduced to less than 25 mph. This allows bikers and walkers ready access to jobs, housing, and stores in a neighborhood, with fewer dangerous interactions with traffic, while still allowing traffic trips.

## Step 2: Transform the station area into a walkable core

The next step is to transform the station area into a walkable core, instead of acres of parking lots and decks (figure 6.7). Rebuild the parking decks at the periphery of this core walkable area. Commuters will not mind that their walk from parking space to transit door now offers cafés, dry cleaners, stores, restaurants, bars, parks, and other services. Property owners will not mind that their "highest and best use" just increased from single family homes to mixed-use walkable density or greater. Increased property values could allow parking to be undergrounded, if real estate and soils/bedrock permit. Transit to and from that station would gain new users who live over and among

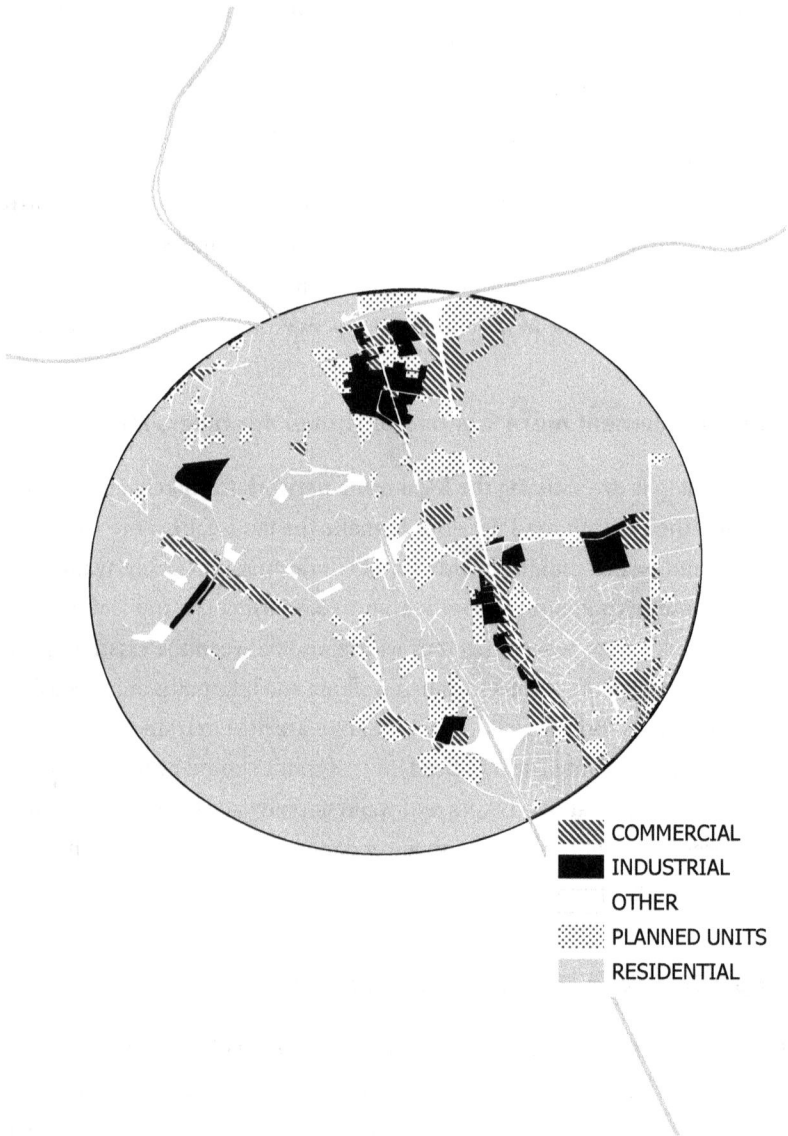

**Figure 6.3. An existing rail transit station and the zoning around it.**

Figure 6.4. An existing rail transit station and the block groups with existing walk-able intensity of at least 14 (HU+Jobs)/acre around it.

**Legend:**
- 0-14 IU
- 14-242 IU

**Figure 6.5.** An existing rail transit station and the road network built for traffic, as well as forest and parkland, for the three-mile radius around the station.

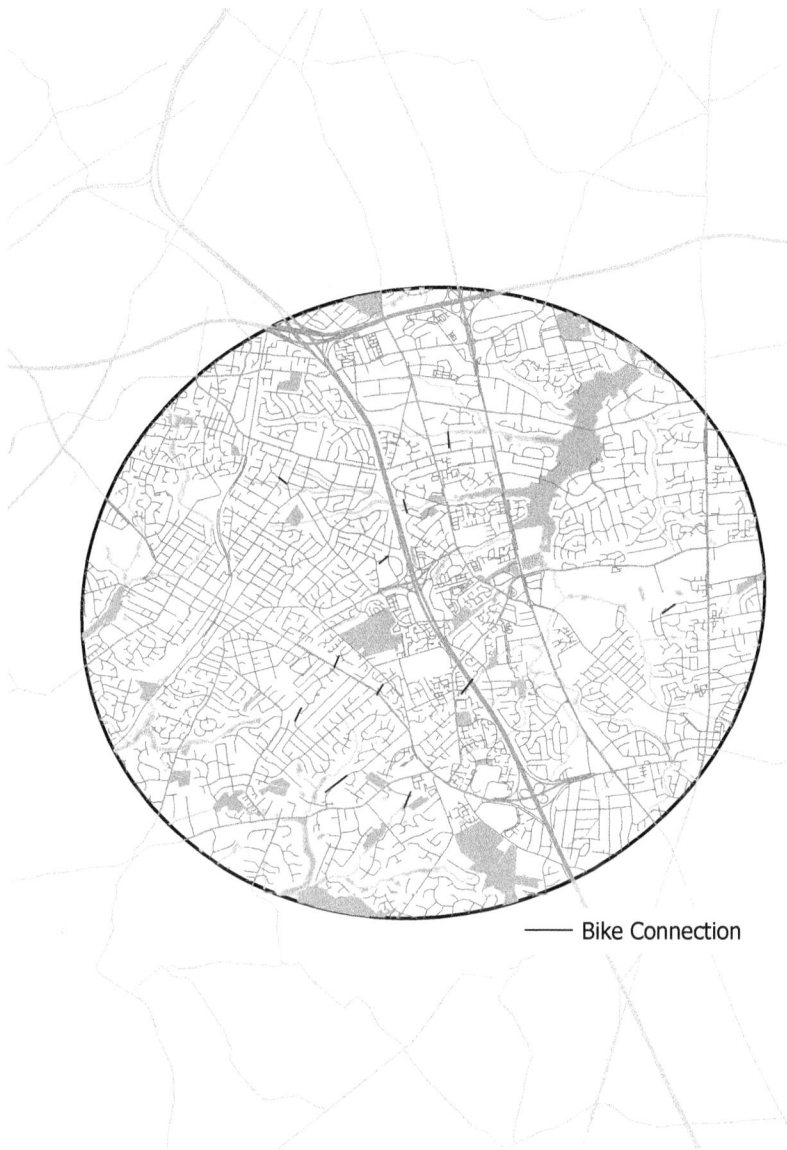

Figure 6.6.  Adding bike and walking connections to the road network.

Mi1 Redeveloped

Figure 6.7. Station area with roadways and paths in immediate station area rebuilt as "transit-oriented development."

those shops, taking the short walk to transit or jobs within the station area. Transit service becomes more efficient as each station becomes both an origin and a destination.

## Step 3: Establish bike boulevards throughout a three-mile station area

The next step is to establish radial bike boulevards connecting the station area to each station (figure 6.8). Bike arterials are still roads, allowing low-speed traffic, but with more space and mobility given to bikes than traffic. Truck delivery along these arterials is limited to early hours. Many retail and commercial establishments already have delivery time restrictions in place, to avoid reduced customer and worker parking capacity. Bike boulevards become the local-scale arterials within the station area, focusing commercial development into "high streets," and allowing delivery of goods by truck, bicycle, and car. The radial arrangement is designed to enable rapid bike access to the station and to the other stations in the metropolitan transit network.

Traffic is not banished from any of these streets, it is just not the mode of choice or the design focus. On a personal and household level, traffic trips are needed only a few times a week for groceries, long errands, and long-distance family journeys. Let parking in the station areas be along the streets, or along alleys and tertiary parking decks designated for the purpose. Most of the space that traffic makes unwalkable and unbikeable is because of the large stopping distance of high-speed, high-momentum vehicles; slow-speed traffic does not cause this problem. Build places where people can use traffic occasionally, transit daily, and bike or walk within them for what they need most often.

Bike boulevards are complete streets, not with an arterial focus or design on traffic, but with biking and walking the primary modes, and traffic secondary. Bike boulevards do not need to be wider than thirty feet to accommodate space for bikers, walkers, and traffic, with on-street parking or loading zones on one side of the street. A typical section would include a six-foot lane for parking and loading, two six-foot lanes for biking separated from traffic by a two-foot buffer, and a ten-foot "skinny" through lane for traffic to move slowly through. The sidewalks are ample, up to sixteen feet wide on

**Figure 6.8. Station area with bike boulevards connecting three-mile radius area with station.**[253]

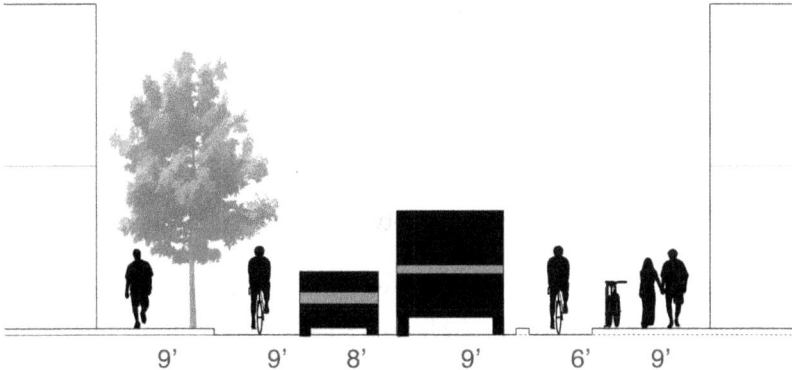

Figure 6.9. Section of human-scaled arterials, with space for walkers, green infrastructure, bikers, on-street parking at market rates, delivery vehicles, and street furniture.

each side, with a tree zone wide enough to accommodate vegetation, street furniture, restaurant or other seating along the curb or next to a store front, depending on the sun and wind orientation of the block (figure 6.9). The focus of the bike boulevards is bikers and walkers, and they are used by people living and working in the station area. The traffic lane and parking acknowledge that many may need a car for some trips, and that most retail, commercial, and industrial goods will come to the station area by traffic. Services like ambulance, fire, and garbage collection will also need large-vehicle access via roads that accommodate these vehicles.

The arterials do not provide two full twelve-foot lanes for two-way traffic. Instead, cars are given ample space to access and leave the transit station area, and they maneuver through it directly, slowly, and prudently. The structure of the street, the sparse signage, the furniture, and people moving around it all signal that traffic is a visitor to, not the focus of the station area.

Bike boulevards interact with the existing arterial network at signalized intersections designed to give bikers and walkers preference when they need to cross out of the station area. Intersection crossings at ground level are preferable to bridges or tunnels; they are cheaper and remind traffic users that they are near a station area rich in walker and biker activity. If a station area develops enough walkable and bikeable intensity, diversity, and prosperity within three miles of its transit station, nearby communities are free to rezone and redevelop in kind. Station areas can be built in a way that takes an approach that we haven't used enough for a century, filling consumer demand for housing, workplaces, and transportation that has been unmet by our traffic-dominated landscape.

## Step 4: Fill in the bikeable grid between the bike boulevards

The final step is to fill in the walkable street grid between the bike boulevards as the market demand for residential, retail, commercial, or industrial uses increases in the station areas (figure 6.10). The existing roads and zoning are made to segregate land uses and be usable mostly by traffic, the upzoned network of walkable/bikeable pathways and arterials around the station area would be built at a different scale: Served by traffic, but not in service of it.

## The benefits

Transit can offer many more trips per lane-hour than traffic (figure 6.9), yet often so many seats go unused (figure 6.11). Does it not make sense to build so that people can use transit more often, for more trips, where it is available? We need to offer Americans a choice between traffic-dependent lifestyles and lifestyles of proximity.

If you add only density of development to a traffic-dependent place, as in a single shopping mall or "walkable" subdivision, you are inviting wider roads and more parking to serve the increased need for traffic. If you add development to a transit-enabled, walkable, and bikable place, you are calling for more development, as more people live, work, shop, learn, and play in that area. The more desirable a walkable, bikeable, and transit-accessible place is, the more it can be built out to serve the needs of residents, workers, and families. Walking, biking, and transit don't need nearly the same space in route or parking as traffic (see chapter 7). The burden of parking and lane space is not needed for places that support other ways of getting around and out.

It is in the interest of passengers, transit agencies, governments, developers—and even traffic users' interest—to let transit station areas be vibrant destinations. Transit passengers can do more without their car, saving on gas while having more to do near their home or work. Governments have higher

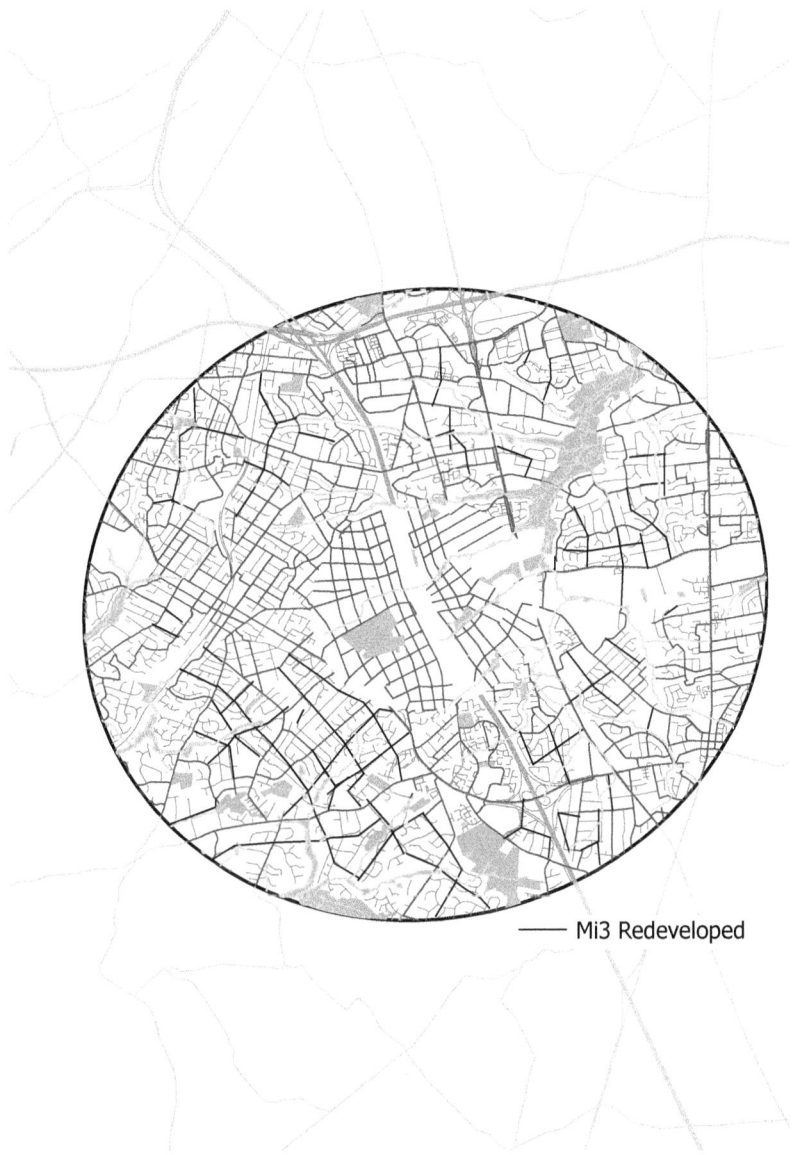

Figure 6.10. Transit station area with network usable by traffic, walkers, and bikers. Walker/biker connections shown in dark black, bike boulevards shown in figure 6.6, revised street grid near transit station shown in figure 6.5, walker/biker connections near transit station shown in figure 6.4.[254]

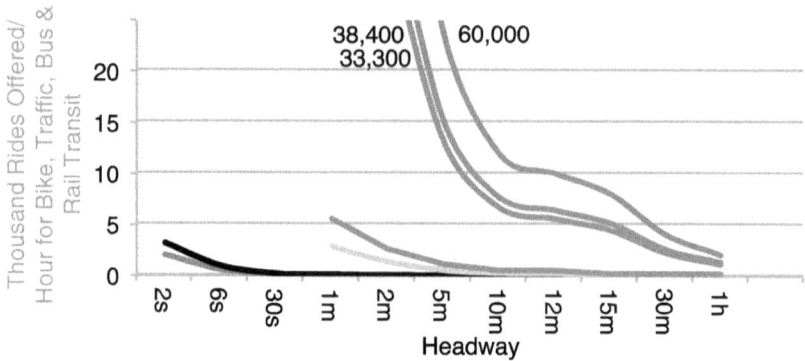

Figure 6.11. Seats offered per hour per lane, by different transportation modes.[255] The four dark gray lines for transit are for bus, light rail, eight-car heavy rail, and commuter rail lines, from left to right.

value land in their jurisdictions with a better tax base, reflecting the use and desirability of the land in station areas. Developers are more likely to recoup their investments, sell their projects, and lease out their space where there is demand. Drivers in traffic on roads surrounding each station area will also benefit from reduced traffic volumes, as the people able to use these walkable, bikeable developments are not on the roads with them. Where people can use transit, bikes, and their feet to get around, they no longer need to be in traffic every day.

Weaving walkable and bikeable paths into road networks also solves one of the greatest costs of traffic dependence in the suburbs: safety of students on the way to and from school. Morning travel congestion is at its peak during the school year as buses and parents ferry kids of all ages from neighborhoods to schools. The areas around schools are some of the most congested as students converge on the school in buses or in their own family cars. Many schools are located on traffic arterials to maximize the access to traffic, which makes getting to and from school even more dangerous for students arriving by foot or bike.

Many students live close enough to school to walk or bike but cannot because of the danger of the surrounding traffic network. Parents and buses have to drive them out of their neighborhoods and onto arterials, even if they live right next to the school, because there are no safe walking or biking paths in their neighborhoods. Fences, woodlots, and hedgerows separate subdivisions and even adjacent streets from one another. Easements (limited public rights to use property) for walk and bike paths through and between properties always require negotiation. If the paths shown in figure 6.8 could be built for the national average price for bike lanes, the whole network shown above could be built for under $10 million. That is about the price of adding one traffic lane to four miles of road.

## Transportation as land use choice

Transportation is a land use choice. How we choose to use our land to live and work determines our lifestyles. We can choose to offer lifestyles of proximity to more Americans, or we can restrict choice as we have done for the last century. Land use choice is about the opportunity of a piece of land—something I explore further in the next chapter. The analysis in this chapter shows the type of land use choices we can make that enable biking and walkability to compete with and outperform traffic as a transportation mode.

# 7

# Opportunity Cost

Our choice of transportation mode is also about how many things are certain distances from us and how much will be involved in using different modes to travel from our origins to our destination. We never drive, bike, or take transit from our bedroom to our bathroom. We choose to live near or far from things, and we choose the densities that are offered to us in our metropolitan landscapes at the prices we can afford. Our ability and choice to take a job far away is defined by the transportation modes we take, and how good the infrastructure is for the modes. Taking a job nearby is a choice many of us in the United States cannot make, as all the jobs near our homes do not pay well enough to afford our housing. Over the last century, we have zoned and built an American landscape that assumes limitless distance and costless travel. This book proposes a new use of the old zoning tools to offer plenty of development where it makes sense—next to existing transit stations—to renovate American development.

In this chapter, we compare the capacity, energy, space, financing, and danger of different modes. Toward the end of the chapter, we consider how these would vary nationwide with different land use/transportation planning scenarios. All the scenarios we consider relate only to the areas within three miles of existing transit stations.

## Capacity

How many passengers can each mode deliver to a place per hour, per lane? The lane capacity of a mode is determined by the number of passengers per vehicle and how many vehicles can fit in a lane in an hour. How many vehicles

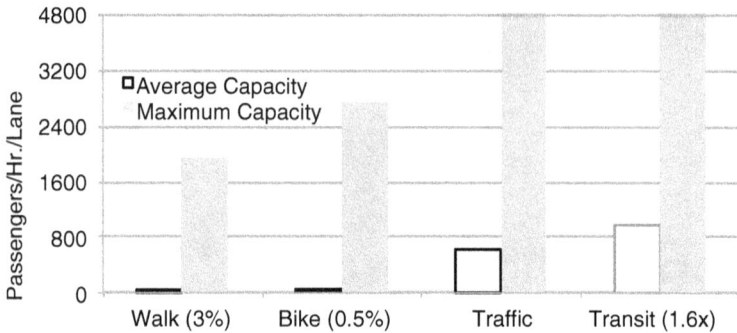

Figure 7.1. Maximum and average capacity of each mode, in passengers per lane per hour. Percentages and proportions shown are in relation to average traffic capacity.

can fit in a lane is determined by the speed of the vehicles and the stopping distance needed to avoid a collision. The faster a vehicle is moving, the farther apart it must be from the vehicles around it, for safety's sake. This is not a regulated speed, but a commonsense calculation we all do when walking, biking, or riding in traffic. In contrast, rail transit stopping distance is not a matter of driver judgment, but block signaling, safety regulations, and engineering. Comparisons of average and maximum lane capacity by mode are shown in figure 7.1, above. To simplify for rail transit, I used the one-minute stopping distance at average speed as the minimum allowable spacing between trains.

No one enjoys traveling at maximum lane capacity. Walkers are crowded and bikers are in imminent danger of wrecking but must keep moving for fear of falling over; traffic is braking and accelerating at a near standstill; transit vehicles are packed with standees, as are the stations. Doors are limited resources on transit, and passengers crowding near the doors makes it harder for other passengers to board or alight. For traffic, the capacity of a lane is usually two thousand vehicles per hour. Drivers are not willing to travel any more densely than that. Walkers and bikers have similar lane capacities, based on their own considerations of speed and stopping distance. Bikers and walkers worry more about traffic than about other bikers and walkers. Running into a person on the walkway or bikeway is not nearly as deadly as when they are in a car in traffic on the roadway.

The design load of a transportation mode is often much less than its capacity. The capacity is the maximum amount of people it could carry, whereas the design load is the expected amount of people it will actually carry. As traffic gets close to capacity, such as during rush hour, it becomes unpleasant and sluggish. As transit, bikeways, or walkways get close to design load, their passengers still get where they are going with a minimum of fuss. Starting and stopping is easier in the case of bikes and walking. Most walkers and bikers are not running far below their average speed at capacity and are therefore less frustrated in their timing. On transit, full vehicles move as fast as empty ones, with some delays for longer dwell times at stations. Even if you have to stand in the aisle cheek by jowl with other people, you are still getting there on time.

The difference between capacity and design load comes up often in the choice between traffic and transit projects. Average transit capacity is often compared with maximum traffic capacity, making the contrast stark and obvious, and supporting the demand for more traffic lanes. Comparing transit and traffic fairly, the maximum capacity of one transit route could carry two million people per day on just two tracks (48,000 people per hour per lane [phpl] x 1 lane for each direction x 24 hours), while it would require eight lanes of traffic (3,200 phpl x 4 lanes for each direction x 24 hours) to carry the same passengers.

## Energy

How much energy does each mode consume to move one passenger one mile?[256] Biking outperforms all other modes and offers a superior return on passenger miles traveled for the energy consumed (figure 7.2). Transit offers the next greatest return, followed by walking; traffic offers the least efficient performance for the energy consumed.

Energy use in transportation increases with stop-and-go traffic. A car or train moving uninterrupted at 60 mph is much more efficient than in stop-and-go travel. Transit's need to stop at stations to pick up and discharge passengers uses a lot of energy. Avoiding the fuel inefficiency of stop-and-go traffic is a historical justification for streamlining the flow of traffic in cities. Traffic congestion becomes more than just an inconvenience, it becomes an efficiency and air quality issue. Hybrid cars, with regenerative braking, can

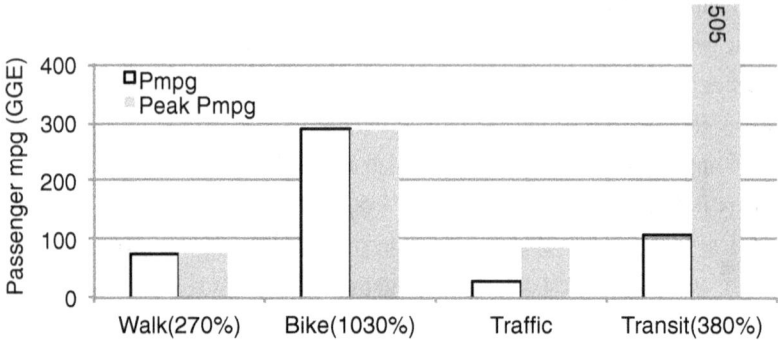

**Figure 7.2. Passenger Miles Per Gallon gasoline equivalents (PMPG), by mode. A Gallon Gasoline Equivalent (GGE) = 114,000 BTU[257] or 33.41 kW-h.[258] Percentages shown next to mode names are average (not peak) modal PMPG in relation to average traffic PMPG, e.g., you get 1030% more miles per gallon on a bike than in traffic.**

increase their efficiency under stop-and-go conditions, but they perform less efficiently on the open highway.[259]

We can also look at energy efficiency in terms of the carbon footprint of each mode. The carbon footprint is a function of the emissions required to convert energy to motion. For biking and walking, this is not so much exhaust as respiration and is directly related to the efficiency of their transportation. Biking offers a mechanical advantage over walking, and as shown above in figure 7.2, is the most energy-efficient mode.

Comparing pounds of carbon emissions per passenger mile, traffic is the least energy efficient, outperformed by walking, biking, and transit in increasing order of efficiency. Multiplying the per-mile emissions by the average trip length of each mode shows the average per-trip emissions (figure 7.3). Per trip, walking is the least carbon-emitting, with biking a close second (even though biking emits less per mile, average bike trips are longer than average walks).

We can also use these metrics to assess tradeoffs in potential carbon emissions. If we somehow replaced all our traffic trips with walking, biking, or transit, we would produce only 3 percent, 5 percent, or 13 percent of today's traffic emissions. Rail transit would emit even less carbon if electricity were generated by wind, solar, hydro, or nuclear, instead of coal, gas, or oil. Carbon-based energy is actually steam energy, with 70 percent of the energy we mine or pump lost to heat boiling water to turn a turbine. America will

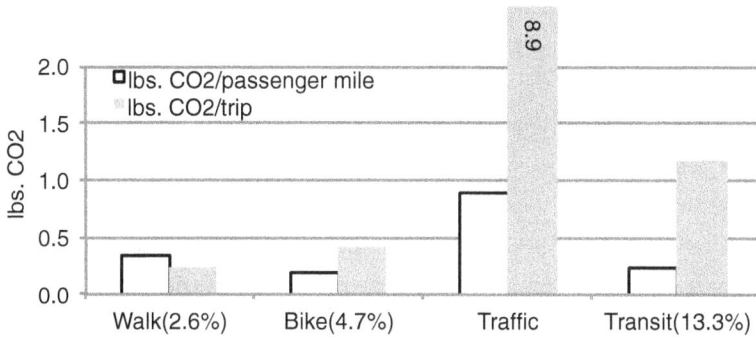

Figure 7.3. Pounds of carbon emissions per passenger, by mode. Percentages shown are in relation to traffic emissions per trip, e.g., the average five-mile transit passenger trip emits 13.3% as much $CO_2$ as the average nine-mile traffic trip.

only need 30 percent of the energy we use now to produce the same electricity when we get over carbon as an energy source.[260]

# Space

Communities have to allocate space for people to get to, from, around, and within places, whatever mode they use. The amount of space allocated varies by mode. Let's look at space impacts and how many people we can move on just one paved acre for each mode.

An acre is 43,560 square feet—about eighteen times the space inside an average single-family home,[261] 680 office cubicles, twenty-one tennis courts,[262] just over nine basketball courts,[263] over five baseball diamonds, or about three-quarters the size of a football field.[264] We can also consider just paving that acre for transportation. An acre is twenty-three dashes on a four-lane highway, about 270 parking spaces, or only thirteen dashes of a six-lane arterial with a median.

Where land is valuable, we care more about what to do with that acre and are less likely to just pave it over. One acre of land can cost as much as $600,000,000[265] in Manhattan, or as little as $1 in Alaskan permafrost. The most valuable places attract the most people and must provide the most infrastructure for transportation. The land that transportation requires varies by mode. Unlike real estate developed as buildings for jobs, services, or housing, land lost to transportation right of way cannot pay for itself. Because land is worth something, the most valuable places are often served by excellent transportation, like high-frequency transit or wide sidewalks that can move more people in less space.

Traffic takes up more space than just the vehicle itself. The speed and stopping distance of vehicles determines lane capacity and space requirements. The bare minimum distance between cars in traffic must permit reaction and braking time: two seconds. In two seconds, a car traveling the national average speed of 31 mph travels ninety-one feet. In the same two seconds, a walker travels nine feet. If you look at any transportation mode, the distance between vehicles will vary by speed, but will almost never be less than the two-second stopping distance for their speed.

Speed also affects lane width. Lane widths are wider than vehicle widths to allow drivers a margin of error before smacking into the curb or each other. While a car is six feet wide and a truck is eight feet wide, traffic lanes are almost all twelve feet wide. Drivers can see how much leeway they have and adjust their speed accordingly. A driver in a ten-foot lane would naturally go slower. The road's capacity would not be any less, since slower cars can follow each other more closely. The width of a walking lane is the width of a wheelchair—under the Americans with Disabilities Act (ADA), about three feet. Most sidewalks are built wider than three feet, to enable passing. At four feet, a bike lane is twice as wide as the average bike, but only a third the width of a traffic lane (figure 7.4). A rail transit lane can be as narrow as the vehicle itself, although trains are usually given a foot of extra berth on both sides for the safety of everything around them.

As with capacity, we are interested in the space per passenger, not per vehicle. The average occupancy of a car, truck, or SUV in traffic has been 1.6 since 1980. The average occupancy for a rail transit vehicle is twenty-seven passengers. Rail transit is usually served by trains of two to ten vehicles and has longer stopping distances than any other mode. For figure 7.5, six-vehicle trains were assumed for the national average between more than fifty rail transit networks.

Figure 7.6 shows the number of users who can be accommodated on one paved acre for each of the four transportation modes. Traffic carries the least number of people in that acre. To express this in everyday terms, more

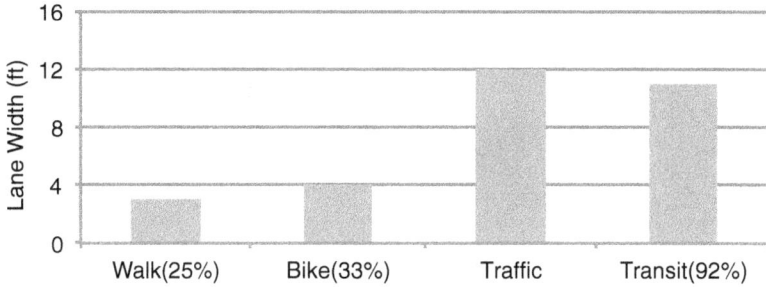

**Figure 7.4. Average lane width by mode. Percentages shown are in relation to traffic lane width, e.g., a three-foot-wide sidewalk "lane" is one quarter the width of a twelve-foot-wide traffic lane.**

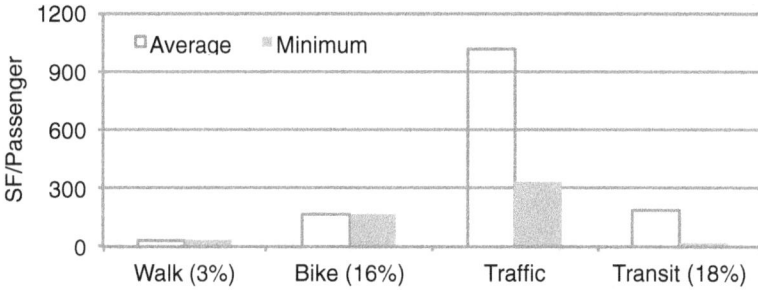

**Figure 7.5. Space per passenger, by mode. Percentages shown are in relation to average traffic space per passenger, e.g., at average vehicle occupancy, with average stopping distance on average lane widths, a transit passenger takes up 18 percent of the space as a passenger in traffic.**

walkers can fit on two nine-foot-wide sidewalks alongside the roadway than can fit in four lanes of traffic or transit as shown.

Figure 7.7 shows space requirements for each mode at average vehicle occupancy (outlined bars) and for maximum vehicle occupancy (gray bars). The factors at the base of each bar are comparisons to maximum capacity of traffic. For example, biking allows six times as many people per acre as traffic. Traffic, even at maximum occupancy of five people per vehicle, is still less space-efficient than any of the other modes. It would be difficult to arrange whole roadways full of this arrangement, as traffic vehicles are filled with people who disagree on their origin, destination, or timing. Transit at maximum occupancy is even more space-efficient than walking.

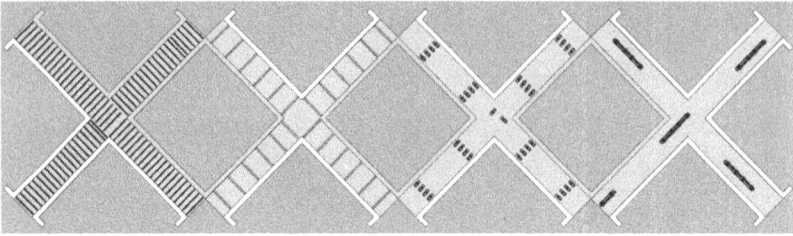

Figure 7.6. Left to right: 1,400 walkers, 290 bikers, fifty-five traffic users, and 230 transit riders, each on four 210-foot-long, forty-eight-foot-wide blocks of paved road, or one paved acre.

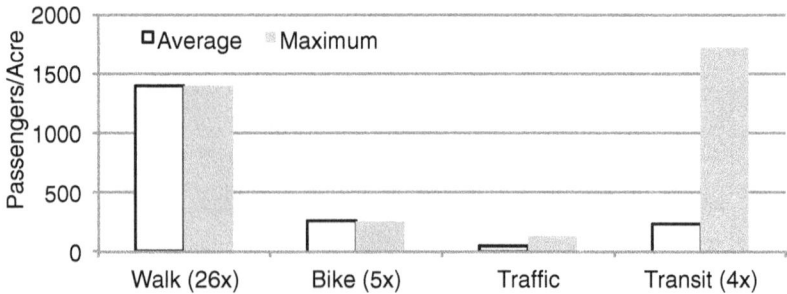

Figure 7.7. Moving passengers per paved acre, shown at average and maximum vehicle occupancy by the gray bars. Proportions shown are in relation to average traffic space per passenger, for example, five times as many moving bikers can fit on an acre of roadway space as traffic passengers (average occupancy of one driver and 0.6 passengers).

These comparisons only account for transportation moving through your community on the way to someplace else. At the beginning and end of every journey, traffic users and bikers must park their vehicles somewhere, whereas transit users and walkers do not impose this burden. Transit station footprints are not essential to the completion of transit journeys. Transit stations can be at, above, or below ground. Walkers walk to their destinations and require only the space needed for their movement. The parking space requirements for bikers are minuscule compared to traffic. An average bike parking space is six to nine square feet. An average car parking space is 162–70 square feet, fifteen to thirty times larger. Figure 7.8 shows the same comparisons as figure 7.7, with the added space requirements of parking.

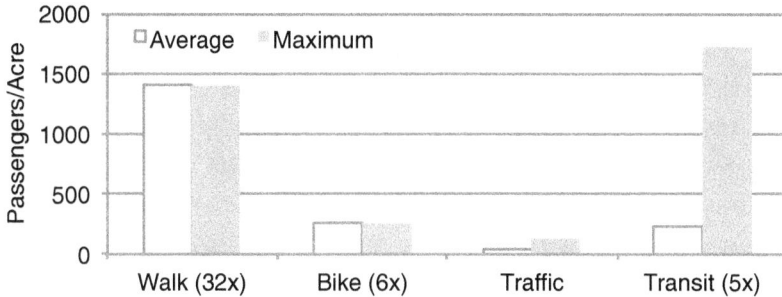

Figure 7.8. Passengers per acre at maximum and average occupancy, including parking. Proportions shown are in relation to average moving and parking space per passenger.

Column graphs do not capture what the space implications of the different modes mean to real people. Bikes and cars in traffic mean different things depending on whether you are sitting in/on them or outside looking at them. A transit passenger moving through has very different impression of a place than a walker or a traffic user. The more people who want to go to a lively place, the more they will demand different ways to get there. If the mode you build for requires large spaces just for circulation and parking, it will deaden the concentrations of uses that made your space so lively before. We should be mindful of what building only for traffic means for our communities.

## Financial cost

Now we consider how much each mode costs, and how we pay for it. The gas tax is the single biggest source of revenue for modes of traffic, transit, biking, and walking, but pays for less than 55 percent of the building and maintenance of roads. The other 45 percent is paid for by local and federal governments taking on more debt. Most of the gas tax is collected and spent by states and localities on their own transportation infrastructure. Federal money is mainly used for large "big vision" projects like the interstate highway system, river bridges, or rail transit extensions.

Who pays for what part of transportation determines whether a transportation mode is considered subsidized or a market good. The road network for traffic is our largest installed infrastructure project, with 8.5 million lane miles over 4.5 million route miles. It costs governments almost $120 billion a year to maintain and expand it. It has cost local, state, and federal governments $8 trillion (as measured in 2020 dollars) over the last century to pave, expand, and maintain it to its current extent. Governments do not own and operate most of the vehicles on these roadways, however. The operating cost of traffic is $1 trillion per year and over $70 trillion over the last sixty years,[266] paid for by all who own a car.

By contrast, large rail transit agencies are responsible for their routes as well as the vehicles, and have to find money to build, maintain, and operate all dedicated railways and vehicles. The transit mandate is to relieve traffic congestion, serve passengers, and concentrate development opportunities around stations. These mixed goals can conflict and are expensive to service all at once. If something goes wrong with transit, passengers and pundits know exactly who to blame—the transit agency. Passenger fares rarely cover the expense of each trip, requiring funding from several levels of government. This is why transit is "subsidized," but roads are "funded," even if from local, state, or federal debt.

The final twist in infrastructure finance is how sidewalks are funded. Most municipalities build sidewalks as a default feature of local roads. The expense of the maintenance of these sidewalks is the responsibility of the adjacent landowner, not the town or city. The presumed beneficiary of sidewalks is the adjacent landowner, not the traveling public. The presumed purpose of a sidewalk is to cross from the home or business to a waiting vehicle, in traffic. Less common bikeways are maintained as part of the public road or the private walkway, depending on which side of the curb they are on.

To compare the costs of the different modes we must consider new capital construction, existing way maintenance, and vehicle operation. Rail transit is much more expensive to build and maintain than any other mode, but it can carry many more people in less space with less energy per passenger, as shown above. Traffic and transit are both expensive and well-documented, while biking and walking costs are harder to discover. Given that the numbers for

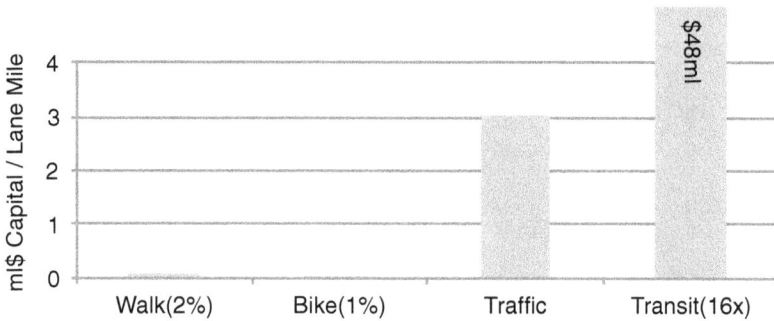

Figure 7.9. Capital cost per new lane mile, by mode, as measured in 2010 dollars. Percentages and proportions shown are in relation to cost of new traffic lane mile. Rail transit is much more expensive than any other mode per lane mile because it requires more secure rights of way and stations to allow passenger boarding and alighting.[267]

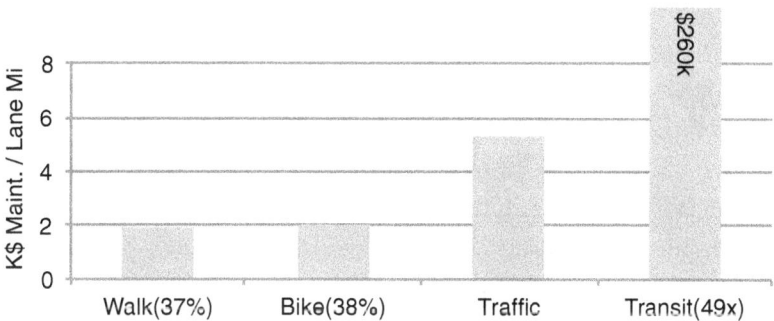

Figure 7.10. Maintenance cost per installed lane mile, by mode, as measured in 2010 dollars. Percentages and proportions shown are in relation to cost per traffic-lane mile. Bikeway and sidewalk maintenance are given as similar costs, as there is no nationally available data on their separate cost of maintenance.

walkways and bikeways are less certain, we can estimate them from specific project budgets to get a rough estimate of costs per lane mile. Given these differences in nationwide documentation between modes, the next three graphs are capital cost per new mile (figure 7.9), maintenance cost per existing mile (figure 7.10), and operating cost per passenger mile traveled (figure 7.11).

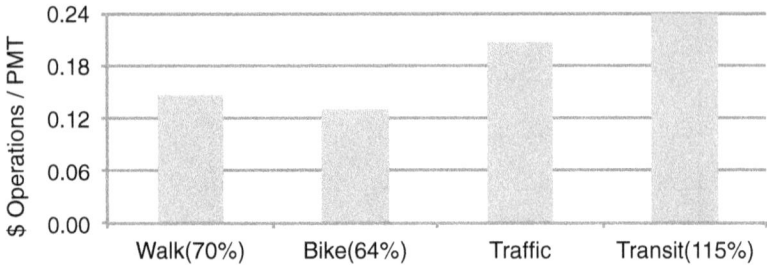

Figure 7.11. Operations cost per passenger mile traveled, by mode, as measured in 2010 dollars. Percentages shown are in relation to cost per traffic-lane mile. Transit operational costs are for the transit agency to carry one passenger one mile, whereas the operational cost for walking, biking, and traffic is for the modal user to travel one mile on the paved surfaces already built for them in shoes, on a bike, or in an automobile.

## Safety

Another way to evaluate the service level of a mode is its safety. What is the risk of injury or death for a trip in each mode? The Department of Transportation's Fatality Accident Reporting System (FARS) reports all fatal accidents on the roads for drivers, passengers, bikers, and walkers. The National Transit Database (NTD) records fatalities and injuries for rail transit. The Bureau of Transportation Statistics (BTS) also records the numbers of injuries on America's roads. By dividing these figures by trip numbers for each mode, we can estimate the chance of mishap for each trip that we take.

Though FARS and BTS data is limited to fatalities and injuries on the road, by traffic, these represent the majority of fatalities and injuries to bikers and walkers in the US. Per trip, bikers experience the highest rates of fatalities and injuries of any mode (figure 7.12). Transit is the safest of all modes. Figure 7.12 shows that walking—considering both risk of injury and death—is only 50 percent more dangerous than traffic. There is greater risk of injury in traffic, but greater risk of death as a walker.

Among the other ways of measuring risk, two others are the chance of collision and perceived chance of accident. Cycling and walking amongst traffic is terrifying in many parts of America, while using traffic is designed to be safe and predictable. Transit, while it is safer by fatalities and injuries, sometimes

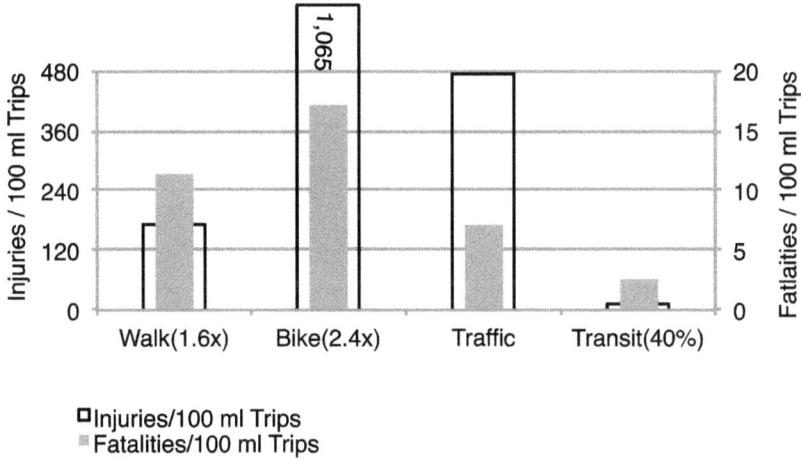

Figure 7.12.  Risk of death or injury per trip by mode. The proportions or percentages given for each mode are the risk of fatalities. The solid gray bars represent the death rates, and the outline bars represent injury rates. Note that the high rates of walker and biker fatalities and injuries are largely due to collisions with traffic.

carries risk of personal crime. The crime rate on transit is lower than for the nation, but it may be the highest-risk time of many riders' days. If transit represents the most dangerous time of many people's day, they are more likely to leave transit for traffic.

After this comparison of several aspects of the four transportation modes, we examine how use of the different modes could affect people's daily commute.

# 8

# One Morning's Commute

**N**ow, let's put the performance metrics from the last chapter to work. More precisely, let's look at the impacts from using alternate modes of getting Americans to work in a typical morning commute. Commuters leave their homes to earn a living, bringing wealth back to their households. Every weekday, over a hundred million Americans commute to work by different modes, on different routes, and over different distances. In 2010, about 134 million workers commuted to work every morning in America, with about six million additional workers staying home for their jobs. The time most of us spend commuting is when the roads are most crowded, traffic is slowest, and when transportation gets the attention of policymakers.

This chapter compares each mode's performance before the proposed land use solutions in chapter 6 and after it, using the metrics from the prior chapter. This chapter compares walking, biking, transit, and traffic by considering the impacts of a typical morning's commute. As of writing, 134 million Americans leave home to go to work. Their average commute distance is ten miles for traffic, five miles for transit, three miles for biking, and a quarter mile for walking. With those commuter numbers and trip distances, we can compare the impacts of commuting in "baseline" conditions with "upzoned" conditions. Baseline conditions use the existing intensities in station areas to calculate the numbers of commuters in station areas (circa 2010, per the EPA Smart Location Database). Upzoned conditions assume that the station areas have been built out to walkable intensities, and the proportional increases in commuters using transit, walking, or biking. As the 134 million commuters are constant between baseline and upzoned scenarios, increases in non-traffic commuter numbers living in station areas will result in decreases in traffic commuter numbers living outside station areas.

Baseline and upzoned scenarios distinguish commuters by commuter mode—traffic, transit, biking, and walking. Each scenario assumes all 134 million commuters are getting to work, but the distance they travel, and their impact, varies by the scenario's arrangement of modes and land use that supports those modes. Each scenario assumes all commuters who have land use that enables that mode will use it for their morning commute.

The "baseline" analysis approximates the status quo and serves as a national baseline of performance, based on existing transportation policy, without making changes to land use in station areas. In order to get an idea of what a shift in transportation policy may get us, this section compares the performance of each mode under the land use solution proposed in chapter 6. The "upzoned" analysis changes the land use intensity and opportunity for transit, walking, and biking commuters in only transit station areas, and shifts commuters to those station areas to take advantage of the non-traffic commutes that become possible.

## Establishing the baseline

The baseline analysis assumes all commuters who live beyond three miles of a transit station commute by traffic, but approximates the number of commuters within walkable and bikeable distances of transit stations using 2010 census block group data to consider numbers for non-traffic commuters. Data on commuter numbers and mode of travel to work is available through the American Community Survey.[268] The analysis compares the following four baseline scenarios, one for each mode (table 8.1):

Right now, about 86 percent of our 134 million commuters get to work in traffic, which is roughly approximated by the alternative mix for the transit and biking scenarios in table 8.1, above.

The impacts of each scenario can be extrapolated from the findings presented in chapter 6. Each mode is presented in terms of its performance relative to traffic. The outline bar shows the impact of traffic for each criterion while the gray bar shows the impact of the alternate mode. Therefore, shorter, colored bars indicate better performance. Again, the mix of commuters defined for each non-traffic (or alternate) mode relies on their

Table 8.1. Baseline scenarios by mode for one morning's commute, using existing station area intensities.

| Mode | Scenario description by mode, Baseline (aka "Before") |
|---|---|
| Traffic | All 134 million commuters use traffic to get to work. |
| Transit | The 36 million commuters currently living within 3 miles of a transit station take transit to work, the other 98 million use traffic. |
| Bike | The 36 million commuters currently living within 3 miles biking distance of transit bike to work, the other 98 million use traffic. |
| Walk | The 4 million commuters currently living within a quarter mile of a transit station walk to work, the other 130 million use traffic. |

proximity to a transit station in the status quo. The changes between transit, biking, and walking within the baseline scenarios all occur within the 1 percent of the country within three miles of a transit station. Indicated by the shorter colored bars, transit and biking outperform traffic in terms of passenger miles traveled, energy efficiency, and in terms of opportunity cost of space (figure 8.1).

The transit scenario outperforms traffic in all aspects except total costs. Transit is much safer than all other modes, due to infrastructure and procedures developed over the last two centuries. The capital cost of a transit lane is about sixteen times as expensive as a traffic lane, a good reason to make sure that transit receives sixteen times as much use as traffic. The transit scenario assumes all commuters use transit for the length of their commutes, whether they live three miles or a quarter mile from the transit station. The comparison shows costs for transit are about as high as traffic for building and maintenance, whereas biking costs less than all other modes due to the low cost of bike lanes.

The biking scenario, like transit, outperforms traffic in all ways except in safety. A primary reason Americans don't bike more is that our safety is not guaranteed or designed for in most American places. In places where bikes do not have to mix with traffic, biking is much safer than traffic.[269]

The walking scenario—limited to the quarter-mile, five-minute-walk radius from each transit station—is nearly identical to the traffic scenario.

## Transit Baseline Scenario

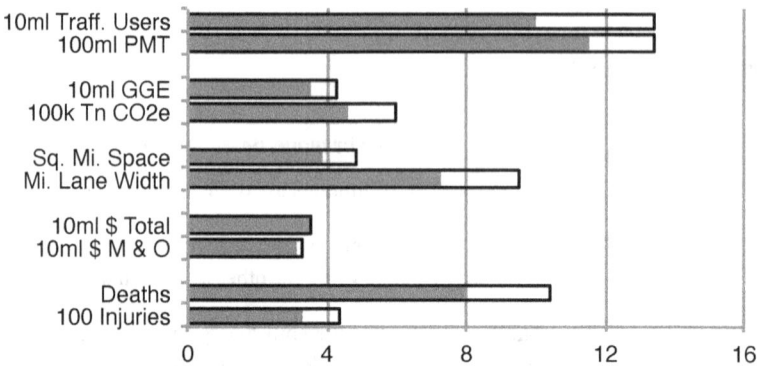

(Bar chart with categories, x-axis from 0 to 16)

- 10ml Traff. Users / 100ml PMT
- 10ml GGE / 100k Tn CO2e
- Sq. Mi. Space / Mi. Lane Width
- 10ml $ Total / 10ml $ M & O
- Deaths / 100 Injuries

## Bike Baseline Scenario

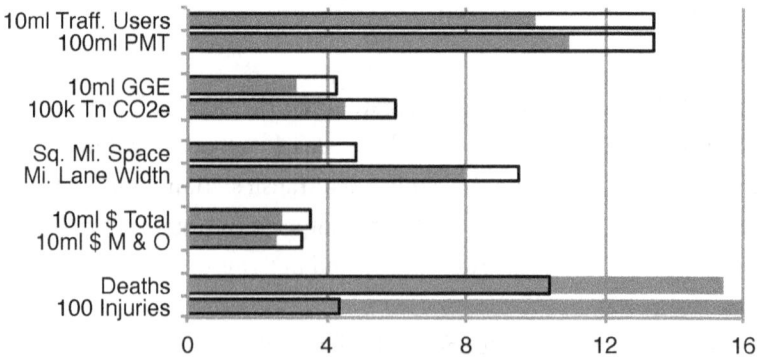

(Bar chart with categories, x-axis from 0 to 16)

- 10ml Traff. Users / 100ml PMT
- 10ml GGE / 100k Tn CO2e
- Sq. Mi. Space / Mi. Lane Width
- 10ml $ Total / 10ml $ M & O
- Deaths / 100 Injuries

## Walk Baseline Scenario

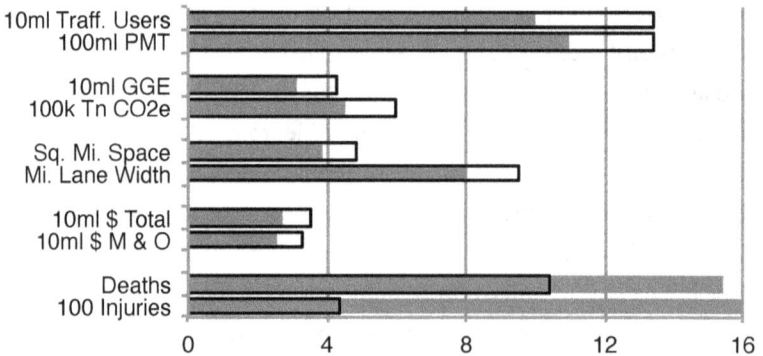

(Bar chart with categories, x-axis from 0 to 16)

- 10ml Traff. Users / 100ml PMT
- 10ml GGE / 100k Tn CO2e
- Sq. Mi. Space / Mi. Lane Width
- 10ml $ Total / 10ml $ M & O
- Deaths / 100 Injuries

Figure 8.1. Baseline scenarios for transit, biking, and walking compared with traffic. The outline part of the bar shows the impact of one morning's commute if all Americans took traffic, and the gray part of the bar shows the impact of Americans near transit stations taking transit, bike, or walk modes for their commutes. See table 8.1 for scenario descriptions.

Because it only affects 0.04 percent of the United States and 3 percent of commuters, the performance improvement from the walking scenario over the traffic scenario is minimal. The area within a quarter mile of every transit station is only 4 percent of the area within three miles, and most of that area is already densely developed. In order to have the same impact as the bike scenario, the quarter miles around every transit station would need to be built up to higher densities than downtown Manhattan or Hong Kong. Compared to the traffic scenario, the walking scenario performs slightly better for commuters in terms of passenger miles traveled and slightly worse in terms of safety. Like biking, walking would be much safer with fewer interactions with traffic.

## Comparing impacts with upzoning land use policy to enable transportation choice

The same framework allows us an objective apples-to-apples basis for comparing the impacts of a morning commute with the solution proposed in chapter 6. That solution specifically targets the area within a twenty-minute-bikeable radius of rail transit stations in the US, and offers the basis for a direct comparison. The difference between the baseline scenarios in table and figure 8.1 and the proposed upzoned scenarios is that development intensities are built out to a walkable intensity of 14 IU within walking (quarter-mile) or biking (three-mile) distances of every transit station in America. How do the baseline scenarios with existing population and job densities compare to upzoned scenarios with walkable intensities?

As commuters shift to increasing transit use, bike use, or walking, and decrease dependence on traffic, national performance of transit and biking become more efficient in terms of passenger miles traveled, energy, and land consumption. How much better depends on the increase in the number of commuters the analysis assumes for each alternate mode. Like the baseline scenarios, the upzoned scenario is still showing the impacts of 134 million commuters. In both the baseline scenarios shown in table 8.1 and upzoned scenarios shown in table 8.2, all commuters living within station areas will use a non-traffic commute mode. The number of commuters from upzoned areas under the scenarios in table 8.2 are calculated by increasing the land use

Table 8.2. Upzoned scenarios by mode for a morning's commute, after land use intensification.

| Mode | Scenario description by mode | |
| --- | --- | --- |
| | Baseline ("Before") | Upzoned land use ("After") |
| Traffic | 134 million commuters use traffic to get to work. | |
| Transit | The 36 million commuters currently living within 3 miles of a transit station take transit to work, the other 98 million use traffic. | The 133 million commuters that could live within 3 miles of a transit station take transit to work, the other 1 million use traffic. |
| Bike | The 36 million commuters currently living within 3 miles biking distance of transit bike to work, the other 98 million use traffic. | The 133 million commuters that could live within 3 miles biking distance of transit bike to work, the other 1 million use traffic. |
| Walk | The 4 million commuters currently living within a quarter mile of a transit station walk to work, the other 130 million use traffic. | The 4 million commuters that could live within a quarter mile of a transit station walk to work, the other 130 million use traffic. |

to a walkable intensity of 14 IU within a five-minute walk or twenty-minute bike ride of every transit station. The numbers of commuters in each scenario are shown in table 8.2.

The scenarios show the number of transit and bike riders who can live and work within three miles of every transit station at minimally walkable intensities. Again, the purpose of this analysis is to understand the possible changes in impacts in each performance category.

Using the ten indicators from the previous chapter on total distance traveled, space usage, energy usage, cost, and danger, we can look at the traffic, transit, bike, and walk scenarios before and after intensification around transit stations. The space and energy impacts of the transit scenario are even lower than the traffic scenario than before, while changing only 1 percent of the American landscape (figure 8.2). The gray bars show the change in impact from the upzoned scenario, while the hollow outline and lighter-colored

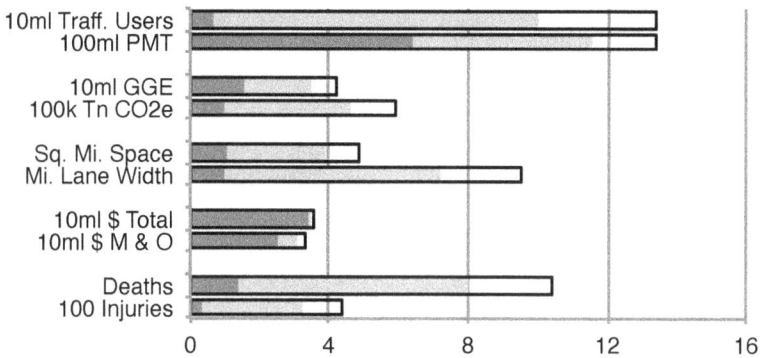

## Transit Upzoned Scenario

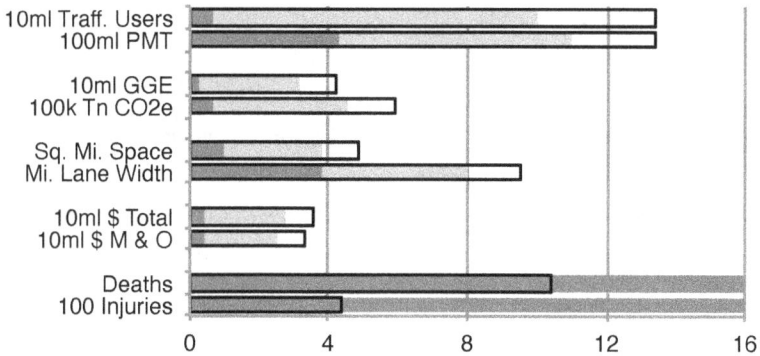

## Bike Upzoned Scenario

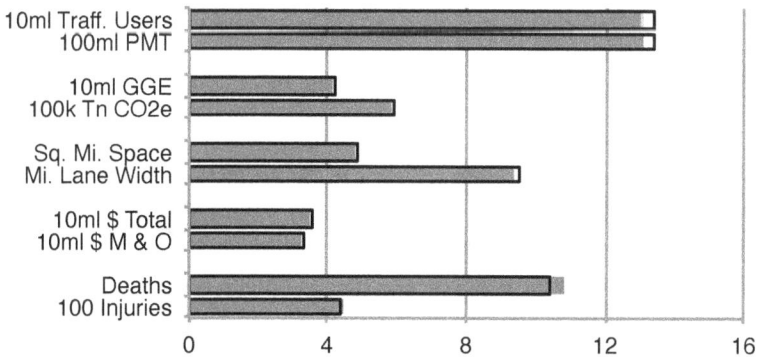

## Walk Upzoned Scenario

Figure 8.2. Comparison of mode performance in baseline and upzoned scenarios. This figure is similar to Figure 8.1, except the light gray bar shows the potential impact of upzoning station areas to bring enough people and businesses to the station areas to enable walkable intensity. See table 8.2 for descriptions of the scenarios.

sections of the bars show the same traffic, transit, bike, and walk scenario impacts from figure 8.1.

The space needs of transit passengers could be even lower than shown, as increased transit use would increase transit occupancy above the current average, further reducing the space needs per passenger. As ridership increases, transit operation and maintenance costs per passenger mile would also decline. Transit, already much safer than traffic per mile and per trip, significantly increases safety with higher ridership.

As with the transit scenario, the bike scenario assumes that the majority of commuters can bike to their work within three miles of transit stations. In this scenario, bike trips connect housing and jobs in the same station areas. With more competitive and diverse land use in just 1 percent of the country, we can move the same number of commuters with greater savings in energy, space, and cost for more passenger miles traveled. Because of its mechanical advantage, biking is more efficient per mile than even walking. Biking is the most efficient mode of commuting, except for well-occupied transit. Transit vehicles are hundreds of times heavier than each of their passengers, while bikes are ten times lighter than their passengers. Biking is less safe than traffic, mostly because of traffic. Most bike injuries and fatalities come from collisions with vehicles in traffic. Transit station areas built at walkable intensities should also be designed for walking and biking safety, which could significantly reduce the danger of biking in station areas. Lower-speed traffic in station areas would improve safety for traffic, as well as for bikers and walkers.[270] The proposed land use changes from chapter 6 enable more accessible, shorter bike trips in a more intense and diverse landscape and allow bikers to not be forced to mingle with traffic to reach their destination. In the future, it may be possible to adjust this measure parametrically using research on fatality reductions in areas where such land use improvements exist.

In the walking scenario, the number of commuters who walk to work increases slightly with the proposed land use changes within a quarter mile of every station. The area within a quarter mile of transit facilities is already mostly built at walking intensities, with not much potential for population growth or development. Many stations are surrounded by parking, which does offer some opportunity for transit-oriented intensities. To increase walking trips to and around transit stations, it is not enough to simply redevelop the "fifteen-minute walk circle" around stations. To provide a critical mass of housing and jobs around each station, the radius of influence around each transit station should be much larger than a quarter mile in order for the transit station area to provide a critical mass of housing and jobs.

The comparisons in this chapter are illustrative. These scenarios are a simplification, as not all residents of station areas commute by transit, biking, or walking, and not all residents outside of station areas commute by traffic.

## Can we do even better?

The area within three miles or even a quarter mile of most transit stations is built for traffic, not walking or biking. Transit already influences land use as it is today, but not as much as it could. We could do better if the people within these walkable and bikeable distances were able to walk and bike to more of their errands, leisure, and workplaces.

Traffic is inferior in many ways to transit, biking, or walking, except for its speed, comfort, and network extent. America's traffic dependence comes at a cost in terms of energy, pollution, space, danger, and budgets. This chapter illustrated the savings possible with alternate transportation choices for these measures with changes to just 1 percent of the country's land. It showed how and where alternative modes can compete with traffic as a transportation choice and the outcomes they offer. Though we compare the impact of 134 million commuters, the likely commuting population in the future will be much larger.

The range, speed, and space needs of traffic make it inimical to the success of the other three modes. Traffic is inherently a hazard to biking and walking so long as there is no protected space for bikers and walkers, and all roads are built for killing speeds. Transit is much safer than any of the other three modes, but its routes and ridership have been hindered by a road network and land use built solely for traffic.

Just as we now arrange highways and parking to build an America for people in cars, we could organize places for the walker and biker, using buildings, streets, and even waterways to revitalize American places. The best places to do this are where the bikers and walkers are the least likely to be cut off

from opportunities around the city, such as within a fifteen-minute journey of a transit station. I have only profiled rail transit and commuters here
because their data was available. Bus stops and stations can and should be
surrounded with transit-oriented walkable intensity, diversity, and design.
Transit-oriented bikeability and walkability will improve all trips at the local and regional scale, without impacting traffic flow in 99 percent of the US.

How do we enable transit, biking, and walking to offer their services to
American transportation without "playing second fiddle" to traffic? The solution proposed here is to bring doors closer to where they can offer metropolitan access via transit, biking, and walking. Let the area around transit stations
develop to their full potential, not as car parks for mode switching, but as livable places around each station, where diversity and vibrancy of uses makes
a car unnecessary for many of people's daily activities.

# 9

# What We Can Get

What can we gain by letting walkable/bikeable development happen near transit stations? The benefits are economic, environmental, and in quality of life.

## Economic benefits

With upzoning and redevelopment of station areas, more people can live and work near transit stations, and more people can use transit, biking, or walking for commuting and errands.

This is a windfall for homeowners and businesses near transit. Land that was previously zoned for traffic dependence and is now upzoned to be transit-supportive would increase their property values as their land could be used for more than single-family homes or commercial lots surrounded by parking. Property owners would be able to leave their buildings in the same form, but would now have the option to rebuild, sell, or subdivide their land for more valuable, mixed uses. This book is not proposing single planned developments, but enabling existing property uses to change from traffic dependent to choice between transportation modes. Land developed for the use of transit, biking, and walking, instead of traffic, could increase in value by tenfold or more. Mixed-use developments are much more valuable to their owners and to their cities per square foot than single-use, lower-density developments required by twentieth-century zoning.

WALKABILITY IS A PROVEN REAL ESTATE PREMIUM IN THE US,[271] NOT ONLY FOR ITS QUALITY but also because of its scarcity.[272] How much more affordable would walkability be if it was commonly available, and not priced as a luxury amenity? Walkability

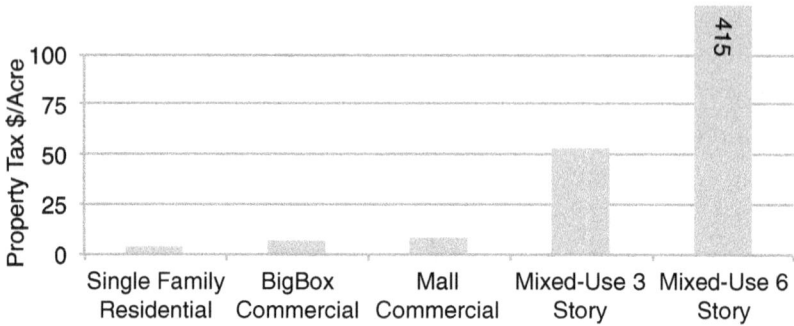

Figure 9.1. Real estate tax return per acre by type of real estate.[274]

should not have to be a matter of high-end design, but of common sense. The intensity and diversity of neighborhoods supportable by transit are not the same as neighborhoods dependent on traffic. The more mixed-use and walkable a development is, the more it is worth per acre. This holds even if the mixed-use development is in a lower-income neighborhood. Studies from Tennessee and Florida by Minicozzi and Katz show that mixed-use developments that enable living, working, shopping, and playing in the same building or subdivision are much more valuable per acre than single-use developments (figure 9.1).[273]

Property values may matter even more to counties, townships, school boards, and any other government that relies on taxes based on property values (although many counties are not financed by property taxes). Communities have fixed areas of land. Communities can undervalue their finite land with traffic-dependent single-family zoning or enable transportation and land use choice to maximize the use and value of their land. Their residents will feel the difference when the community provides more services like better schools, libraries, and emergency responders, as well as more engaged local governments. Communities can choose whether their lands and services are worth a lot or a little, depending on how much land they devote to traffic and parking, and how much they make available for walking and biking.

There is a housing affordability crisis going on right now. The cost of housing is greater than most people's wages can pay for. The stability of homeownership is cut off for most Americans. The reasons for this are several, but a main cause is the requirement that most housing be built as single-family housing at low densities reachable only by traffic. Mixed uses between commercial, residential, and even industrial are simply illegal. Multifamily houses, zero-lot line houses, or even townhouses require lengthy variance processes. All because we've built out our cities and suburbs with the assumption that anyone who can afford a home needs a car to make their living in traffic.

Enforced scarcity caused by single-use zoning codes is an impediment to housing affordability. Parking and setback requirements make jobs and housing sparse and far apart, limiting opportunity for both land and people. Places that are accessible only by traffic tax their residents, workers, and visitors to own and operate vehicles to use traffic, and their property owners to provide parking on their land for those vehicles. We can do better.

# Environmental benefits

### Energy consumption

Another benefit of building better is reduced energy consumption and carbon emissions. The US is directly responsible for 20 percent of the world's greenhouse gas emissions and 27 percent of its petroleum consumption.[275] We have an outsize impact, considering that Americans are 5 percent of the world's population on 6 percent of its land. America's top three emitters of greenhouse gasses are transportation, buildings, and food. The proposal in this book addresses all three. We can reduce our consumption of energy by reducing our need for driving. We reduce our dependence on natural gas and coal for heating and cooling by building closer together with more party walls and multifamily developments within biking distance of transit and walking distance of each other. Building transit station areas for walkability and bikeability would increase the density of customers around supermarkets but would reduce customers' capacity to carry groceries back home. More frequent trips to the supermarket for smaller amounts of food could lead to less waste, as perishables would be harder to buy in bulk.[276] Reduce our energy consumption for traffic, buildings, and our emissions from food, and we make real progress on issues of climate change.

The national energy flow diagrams released annually by Lawrence Livermore National Laboratory indicate the potential of a postcarbon world

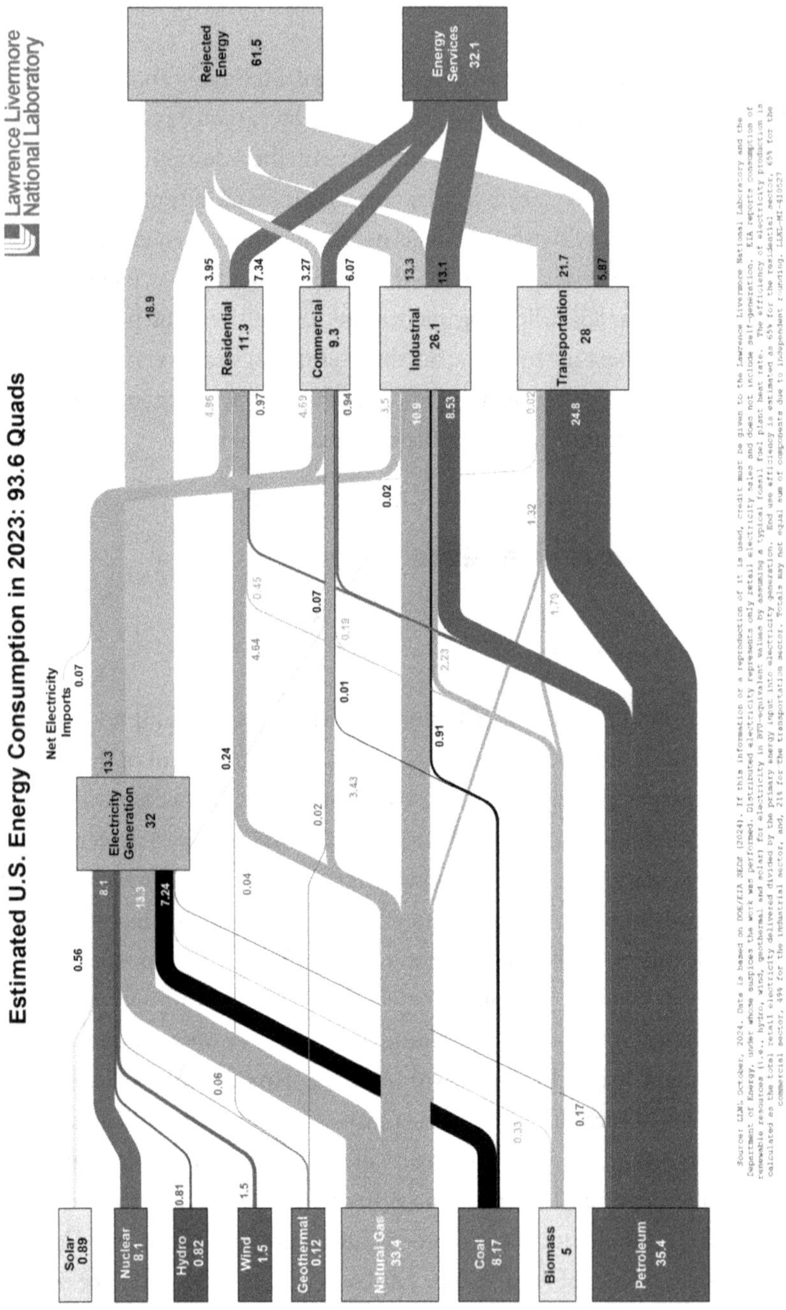

Figure 9.2. Sankey diagram of energy flows from extraction, generation, consumption, and waste heat for the United States, 2023. Provided courtesy of Lawrence Livermore National Laboratories, and part of a series of energy flow diagrams published since the 1950s.[278] This figure appears in color online.

(figure 9.2).[277] The right of the diagram shows rejected energy—heat—wasted during the production of useful power. This waste is appreciable for electricity generation but is highest for transportation. We waste almost four times as much energy as we use to move vehicles in transportation. That does not even account for the inefficiency of moving three hundred pounds of people in three thousand pounds of vehicles. The efficiency of traffic is about 2 percent, by this measure. This book proposes to reduce this impact by letting Americans use traffic for what it is good at, as well as using transit, walking, and biking where they make the most sense. By defining areas for bikeability and walkability, we can reduce the dependence on traffic for a significant portion of the United States' populace and workforce, on 1 percent of our land.

Another energy benefit of proximity between buildings is the advantages of party walls and condo blocks, providing insulation for each other.[279] The heating and cooling demands of the homes in these buildings are greatly reduced, as they are not exposed to outside air on four walls and a roof. In a mixed-use building, most units only have one external wall. Many neighborhoods with walkable intensity are full of single-family homes on small lots. Walkable intensity can also be full of townhomes, homes that are two to four stories above businesses, walkup garden apartments, apartment blocks, or a mixture of these housing types. America has numerous examples of neighborhoods with walkable intensity, and most are not dominated by residential towers or monotonous blocks.

More intensive development allows district heating and cooling. We spend over half of our building energy on heating and cooling, mostly by burning methane[280] for heat and using electricity for air conditioning. District heating and cooling uses a central chilling/heating plant to pipe water for climate control to several nearby buildings, greatly improving efficiency over electricity generation, and even improving efficiency for heating.

## Water

America's century of paving for traffic has damaged America's waters. Parking lots were developed over a century ago to stay dry, jettisoning water to the nearest creek. With every rain, gouts of water are jettisoned from every road, driveway, and from many parking lots directly into the nearest rills, creeks, streams, and rivers. Though stormwater management and Low Impact Development are evolving fields, they are expensive, and they are rarely done in retrofit without redevelopment. Urbanized streams are carved by flashy, drastic flows of stormwater that rise quickly, tear through, and decline almost as quickly with every rainfall. The banks of urban and suburban streams have been transformed into floodways rather than aquatic communities, more like arroyos of the desert southwest than streams in forests or plains. Urban streams have wide and steep banks, with a trickle of flow in the bottom of a wide, rectangular channel. America's worst water pollutant is now sediment scoured from the banks of millions of miles of suburban streams.

One of the greatest benefits of this proposal is for water quality. It reduces the need for metropolitan residents and workers to pave much of their metropolis for the sake of moving and parking cars. Concentrated development around transit stations does require more impervious surface per acre than suburban, traffic-dependent development, but the greater number of people living and working on those acres more than makes up for the local paving and roofing. This can leave more land around the metropolis as farmland, forest, prairie, or wilderness.

Even if everybody is walking and biking, they are doing so on pavement. The roofs of buildings, no matter how tall, are meant to shed water and convey it quickly to the ground and then to streams. The concentration of imperviousness is always greater in developed places, to the detriment of local streams. Stormwater management has developed to reduce the worst impacts of all this pavement, but it is still impervious surface, even with green roofs. Required stormwater management increases the cost and space needs of new development. Like parking, it also makes places less walkable. Bioinfiltration swales and cisterns can enable more compact development, but at greater expense. Soils that are suitable for infiltration are rare. Most are too thick to absorb water as quickly as impervious surfaces will jettison it, or too prone to cavitation and disastrous failure. Stormwater management would be a concern with this plan for infill around transit stations.

By enabling compact developments near transit stations, it is possible to be more mindful of the stormwater impacts that the station area has as a district, not just a collection of upzoned parcels ripe for redevelopment. The per-capita impervious surface is lower in walkable places, but the overall need for impervious roofs, walkways, bikeways, and even roadways is still higher in a walkable/bikeable place than in a new exurb. The total impervious surface needs of a walkable/bikeable station area are still much less than an exurb, as the exurb requires and enables ever more highways and parking to serve its traffic dependence.

There is a problem: what if the concentrated development in station areas just frees up more land for development in nonstation areas? This is like the refrigerator buyer who responds to a more efficient line of refrigerators by buying a bigger refrigerator.[281] As we see from the price of housing, there is an unmet market need for affordable housing and transportation that can provide access to employment and other needs for Americans in their early and mid-careers. A metro with seven hundred thousand people living in walkable station areas and three hundred thousand people living in drivable suburbia is impacting fewer watersheds and streams with less paving than a metro with seven hundred thousand people in suburbia and three hundred thousand in a walkable downtown. This is a proposal for building inward. Right now, America only has a plan to continue building roadways and parking outward. Allowing inward development to station areas can benefit the watersheds serving every metro.

## The city in nature

Can we build cities that coexist with nature as well as function as real estate? If each of us in that city needs 2 percent of an acre paved to use traffic, then likely not. If some of us could need less than 0.2 percent of an acre paved for walking or biking to our destinations, then nature would have more space to be a part of and a complement to real estate. Nature could be an asset, not an afterthought that gets cleared away to provide more parking. Green infrastructure, like street trees, parks, forests, planters, and green roofs improves the ecology of cities and increases property values where they are encouraged and affordable.

You can map prosperous neighborhoods in many cities by how many street trees they have intact. Only the wealthiest households can contemplate the expense of a tree falling on their home. Poorer property owners avoid risk by planting smaller trees and cutting down larger trees as soon as they can afford to. The oldest parts of cities are often the most treeless. Everything is given over to buildings and transportation. The natural and drainage benefits of less impervious surface per capita only benefit forests and streams if there are any street trees, parkland, or undeveloped nature left to provide habitat and filter runoff. Building for walkability, not parking, emphasizes the urban and suburban tree canopy and street trees as an asset to biking and walking. It emphasizes street trees as improving the quality of life in the city, not as fixed-object hazards and nuisances to parked cars. but as shelter, insulation, and nature.

We could have a verdant quality of life in so many more places if we didn't need so much lane space and parking for every vehicle. Transportation occupies space that could be devoted to natural amenities like street trees, pocket parks, and city parks. How we choose to use our space reflects our priorities.

Development around transit stations does not need to be entirely developed real estate and landscaped nature. Places can both function as real estate and coexist with nature by using the lay of the land. Trails next to creeks or ponds let people visit and know their local waters with shared-use trails athwart the street grid. A complete tree canopy over narrow streets and hedgerows along sidewalks lets nature inhabit cities in ways not allowed by parking lots or wide traffic lanes.

Green urbanism should offer places within walking distance of each other. Street trees, pocket parks, and curb filtration swales add to the comfort of walking. Large stormwater ponds built to treat the runoff of larger parking lots, or of large remnant woodlots only space landscape and walking distances out further. People want to walk and bike to things to do. Reduce the need for paved surface and treatment of runoff by reducing the use of traffic in walkable places. Street trees, a strong urban canopy, and riparian forests with trail networks connecting the terrestrial road and path network are one way to do this.

Concentrating development in areas around transit also reduces impacts on nature in other parts of the region. Major roads fragment nature into isolated blocks. Reducing the need for major roads in these areas can knit more of nature's connection back into them. The diversity of any natural community in any developed landscape is determined by how many species can cross the roads that surround each fragment, and how good they are at hiding from predators along the edges. Transforming transit station areas into walkable/

bikeable places can improve the quality and capacity of nature at both the local and regional scale.

## Quality of life

This is a broad proposal to allow better places in America. Better urban design books have already been written and are being written in response to changes in markets, technologies, and needs—several are referenced in the bibliography. We know how to welcome walkers and bikers into streets. We have done a poor job for the last century welcoming walkers and bikers by default, because we have been building for traffic as the only useful mode of transportation. Use America's space better and renew freedom to move as we choose. Not as we must—by traffic—but as we can, on foot, bike, transit, or traffic. Make our places better to gather, interact, create, and reestablish America's freedom of choice.

The varied streets of old town Alexandria, Virginia, the backyard gardens of the French Quarter in New Orleans, Louisiana, and the streets of Manhattan, New York, are all examples of the walkable city. That walkers use places that are inviting to them is not a mystery. The mystery is why we don't build more places like them, even though we know how to do it. The mystery is why we made it illegal.

TRAFFIC IS WELL-SERVED BY SUPERBLOCKS OF PARKING SURROUNDING BUILDINGS. Walkers and bikers are served by a mix of different types of land use and different building types within walking and biking distance of each other, connected by paths safe from traffic. Transit is well served by density and diversity of land use types—origins and destinations—within walking and biking distance of every bus stop and rail station. Land use, building type, cost, and tenure type can all be mixed within a neighborhood to maximize the opportunities for people to get what they need from a neighborhood on foot or bike.[282] The default zoning mode for decades has been to segregate by cost and tenure type among subdivisions, and even among whole school districts.[283] We weaken ourselves and our communities when we isolate ourselves from people who are different from us. Traffic enables isolation; walking, biking, and transit all enable us to be Americans in community again.

**STRIP MALLS ARE A SEA OF PARKING IN FRONT OF A ROW OF STORES PARALLEL TO THE** road because that is what traffic needs to operate: parking. The parking is in front because the first thing drivers want to know when they reach their destination is "Is there parking for my car and can I complete my journey?". Conversely, walkers and bikers only need safer ways to get where they are going within walking or biking distance. Walkers don't need parking, and bike parking can be in back to keep the storefronts next to walkways and street life.[284] Walkable streets make the buildings and the streets feel like they are part of an integrated whole, with safe access to all parts of the street between buildings. In most American streets, roads, or "stroads," walkers are lost in a field of pavement and lawn, scurrying along the baseboards of buildings and in mortal danger if forced into the middle of the right of way along broad street crossings.

Roadways, land use, and buildings have been designed—meticulously and expertly—for the optimization of traffic flow. The city made to serve traffic is hierarchical, with highways linking tiered networks of arterials, collectors, and local roads so that traffic can reach any address on the network with a calm beginning and end to every traffic trip. Walkability, bikeability, and urbanity suffer because they are considered subordinate to traffic-oriented design. Good urban design works at many scales including street, neighborhood, and wider transportation. A city made for walking and biking is more a collection of linked neighborhoods than a regional network of highways. Biking from one end of the city to the other should feel continuously safe, even while passing through dozens of neighborhoods.[285]

**TRAFFIC OFFERS METROPOLITAN ACCESS BY CONNECTING EVERY ROAD TO EVERY OTHER** road. Transit offers metropolitan access by building up every stop and station to serve and be served by the quality of transit to that place. Bus stops should have hundreds of jobs and housing within walking or biking distance of them, just as heavy rail stops should have thousands. Suburban highway interchanges already have thousands of jobs and housing within driving distance of exits. What could we have if transit formed the desire lines between places, instead of traffic?

Traffic is a known weapon against walkers, used in dozens of terror attacks in the last decade.[286] More than that, traffic is a daily attack on our sense of safety, well-being, and control, even if we are in the driver's seat.[287] As detailed in this book, paying for traffic is an obligation—a tax—that households, businesses, towns, cities, regions, and states have taken up because we have foreclosed all other choices for most of America. America took up traffic with the promise of freedom from monopoly control by trolley and rail barons. It is time for a promise of freedom from dependence on traffic.

**AS RICK HALL SAID, "THE GUY WITH THE SIMPLE JOB ALWAYS WINS," WHAT ARE THE** simple rules to enable walking, biking, and transit use as relevant modes for transportation in America, or what could they be? This book outlined a dozen rules at three scales for walkable communities. Traffic engineering followed just four rules for most of the last century.[288] Traffic "level of service" is a simple measure of congestion. The object of traffic engineering is enabling cars to move as fast as possible. Walking, biking, and transit levels of service are complex calibrated polynomial equations, far from intuitive or usable by most people.[289] The continued need for expert judgment in anything but traffic means that traffic is still going to win. To win, human-scaled transportation must be simplified. Jan Gehl's principle for walkable urbanism is best: "Be sweet to them (walkers and bikers) and they will be sweet to you."[290]

A walkable community is a community that sees each other. Not just walking their dogs, or playing with their kids in the nearest park, but for everyday trips like shopping or commuting. We know our neighbors when we see our neighbors. Housing and workplaces are built to be safe, insular places, alien to all but the family and friends we invite in. Traffic exacerbates this isolation by allowing each of us to dart from our carport or front door to our parking space, to get away with minimal eye contact with neighbors or coworkers. The walkable neighborhood invites us to run our errands and make our journeys in public—to be in conversation with our communities.

**THE INVITATION TO MEET EACH OTHER WALKING AND BIKING IS ALSO AN INVITATION** to know each other, to care for each other, to help each other. This includes seeing, talking to, and helping the elderly among us, and watching over and caring for close neighbors' kids. This asks for changes to the boundaries we have from our neighbors in drivable suburbia. This relaxation of boundaries and barriers among neighbors is not for everyone, but there is unmet demand from households that would like more choice and more connection in their communities.

People choose to live in the neighborhoods that they do for myriad reasons. Some settled for what they could afford or for a reasonable commuting distance. Others compromised with their spouse's or kid's needs. Others might prioritize proximity to parents, friends, or churches. Another reason, relevant to this book, is transportation choice. Every new neighborhood is required to be built for traffic access. The few that are built with transit or walkable access are naturally attractive to those who value those things, and they sell for a premium.[291] Walkable places in proximity to transit will fill a market need for lives lived without the obligation of traffic.

**THIS PROPOSAL IS MEANT TO BE MARKET-SUCCESSFUL. IF NEIGHBORHOODS, CITIES,** transit systems, and metropoles that upzone, rebuild, and infill around stations do better in their real estate, jobs, or governance than traffic-dependent places, other places are free to imitate them. This book is an invitation to other parts of the country to build for transportation and land use choice, instead of "one mode fits all" traffic. This proposal will succeed when job creation, livability, and economic competitiveness happens in enough connected, transit-oriented places that other places start to ask how they can use their transit resources to connect a network of walkable/bikeable transit station areas.

## Market

Zoning policy can be changed for the places that we want, but even the best designs depend on a lucrative real estate market to get built. We do not have to accept traffic dependence as our inevitable fate, but we do have to move at the pace of the market. A vibrant real estate market is a greater force for transformative change in development than the best designs. No plan gets built and succeeds without a strong market for its uses. Much of the walkable development that got built in the first decade of the twenty-first century was built under a speculative bubble of easy money. Lucrative real estate markets are also not inherently innovative. Why tinker with what works? No planning, development, or transportation paradigm can overcome a poor economy, but it can help set the stage for a better economy. Many of the largest opportunity sites in American cities are in post-industrial areas with the lowest population and property values.

A lesson from the Great Recession was that we need a strong real estate market to try building interesting things. When the real estate market revived, we kept some lessons and discarded others. A lesson from the COVID-19 pandemic was that the commute was not essential to work. As the pandemic subsides, a new home-based landscape emerges for many Americans, a perfect case for walkability. This book is a proposal for a better set of rules. A lesson from our recovery from the COVID pandemic is that we still need to enable affordable and walkable places for Americans to live and work, instead of forcing them to drive until they qualify for places to live.

# 10

# Conclusion

With COVID-19 lockdowns, fear of transit, and working from home, I wondered if there was an audience for this book. I reminded myself that most of America still lacks transportation choice, still needs to match modal service to land use, and is still grasping for sustainable transportation policy.

I wrote this book to understand where traffic was in relation to walking, biking, and transit, to settle some arguments about smart growth versus traffic advocacy. My analysis was straightforward. I didn't want to find hidden answers, but answers that were hiding in plain sight. After the history of how we got to our current conditions, I looked at how traffic has been for America, good and bad. I presented national evidence at a local scale, to understand the issue at the scales of traffic, biking, walking, and transit. I compared America's transportation modes to see if they could *ever* outcompete traffic. If they could, were there particular places where alternative modes worked best? And what would such places look like? I proposed a solution that everyone can understand, based on objective facts and figures. My aim was to contribute to a productive and evidence-based debate on America's transportation future—one focused on examining choices and understanding places, not dueling anecdotes.

The problems and solutions around traffic do not solve themselves. Though expensive, traffic remains the default pathway to prosperity in most American regions. We lift ourselves out of poverty on foot, on bikes, and on transit, but most of American prosperity is only available through traffic. We have been willing to sacrifice hundreds of thousands of American lives and spend billions of dollars every year for this model. We need better choices.

Traffic has changed in response to two energy shocks and two recessions in America in this century. America's invasion of Iraq in 2003 disrupted global energy prices for several years, ushering in the first hybrid vehicle boom. The Great Recession of 2007–09 and its faltering recovery saw a significant decline in driving. Many Americans could not wait to get back to growth and unlimited traffic. With the global COVID-19 pandemic, lockdowns and supply chain disruptions brought another major recession and another decline in traffic. We spent the years since learning about local modes like walking, biking, and working from home. America will use transportation, space, or commuting differently in the coming years. With pandemic supply chain issues and oil disruptions from Russia's invasion of Ukraine, traffic is becoming more expensive to operate than before.[292] Traffic is beginning to become a luxury good.

The conversation for many transportation policymakers is still only about traffic versus transit. Zoning developed and continues to serve the separation of land uses into a landscape only navigable by traffic. Biking and walking are not seen as realistic modes for anything but personal fitness. Transit is concentrated traffic congestion, nothing more. Accessibility appears in some conversations, but it is easy for people to look at accessibility and dismiss it as not about transportation at all. Proximity is barely considered a serious form of transportation, even where it could fulfill the purpose of transportation, that is, getting there on time. Biking and walking are important transportation modes for the tens of millions of Americans who cannot afford a car, but they must move on networks in landscapes built primarily for traffic. What most walkers and bikers in America are working toward is a time when they can afford to join traffic in cars of their own—to never have to bike or walk again.

**BECAUSE AMERICA DOESN'T OFFER TRANSPORTATION CHOICES, IT ONLY OFFERS TRAFFIC.**

**AMERICA HAS NOT YET DEFINED A POLICY FOR TRANSIT, BIKING, AND WALKING AS** transportation choices on their own terms. They remain stretched to traffic's scale, planned and financed in relation to traffic. Traffic's dominance has come at a monumental cost. Apart from the economic and environmental costs, traffic is still the leading accidental cause of death for teenaged and young adult Americans. Motorized traffic was an excellent update to the horse a century ago. It is time for an update to traffic.

The solution proposed in this book is to let walking, biking, and transit affect land use and the development of our cities as much as traffic. Give transit passengers, bikers, and walkers places that are theirs by right. Use existing transit networks to connect transit station areas into an alternative regional transportation network. Traffic needs billions of parking spaces on private property to keep traffic from dying of arteriosclerosis. Transit would be equally served by diverse, prosperous, and walkable places within walking and biking distance of every transit station. There is no reason a transit passenger needs to drive to or from a transit station, if they can reach their origin and destinations on foot or by bike. Walking and biking are most useful when there are plenty of things reachable in a safe walk or bike trip. We can achieve this synergy through land use changes to just 1 percent of the total land area today.

The 1 percent of America's land in rail transit station areas is just a pilot proposal. We haven't invested nearly as much in transit, biking, and walking as traffic in the last century. Only traffic had a land use and property mandate for parking and "right of way." This book proposes upzoning and redeveloping bikeable areas around four thousand rail transit stations in twenty-nine metros. If that works, there is no reason this can't be applied to the eight hundred bus networks in the US as well. Just as it would be a waste to buy a car for tens of thousands of dollars and not drive it, it has been a waste to buy transit networks for hundreds of millions of dollars and not use them. Let transit, walking, and biking have a place in the American landscape.

The quality of transit determines the extent of the influence around each station. A heavy rail transit line has as much passenger capacity as fifteen traffic lanes, so it should influence the walkability and bikeability of a wide area (i.e., three miles, as proposed in this book). A suburban bus stop offers lower service to each stop—perhaps upzoning of five to twenty-five acres around each bus stop to enable 14 HU+jobs/acre of mixed-use development. Only 8 percent of the land within three miles of rail transit stations is developed at minimal walkable intensity of 14 IU.[293]

These proposed changes to our transportation and land use do not have to happen all at once. Property owners can profit as their property becomes more valuable with more options for building types and access. If a station area, transit network, or metro is made more economically successful by diversifying its transportation, housing and land use options, other places are free to upzone and redesign their station areas. This is not a one-size-fits-all proposal, like parking minimums or roadway design standards. It is an invitation to allow choice. If it works, try more choice. Early failures will invite us to do better. If a single subdivision near a transit station implements this and profits by it, other subdivisions are free to try it as well. Early successes will call us to try more.

The pandemic was a challenge to several points in this book. America discovered that jobs that involved working at a networked computer didn't inherently need a commute between home and work. Traffic plummeted in the spring of 2020. As we have become acclimated to the pandemic, many Americans are still working from home. Roads that were once oversized for twenty-two hours of the day are now overbuilt for twenty-four hours of the day. Drivers in traffic, as well as bikers and walkers around traffic, are at greater risk around the clock, as more traffic is moving unobstructed at greater speeds.

The pandemic's effect on transportation, work, and commercial office space illustrated another point: America's energy needs have peaked. America's overall energy consumption declined over 7 percent from 2019 to 2020, mostly due to lockdowns and work-from-home policies. It has still not returned to its 2019 peak. America's total energy consumption peaked around 1998–2000[294] and has not significantly increased since then, even as our population has increased 17 percent by almost 50 million. This is largely the effect of the internet replacing the need for trips.

Even though we have built out millions of miles of roadways for traffic across the US, we are not bound to use traffic as our sole mode of transportation. Technology often becomes most advanced before it becomes obsolete. For example, the world's current consumption of silver is down from its peak in the early 2000s, when Kodak's exit from the film camera market after digital photography made film photography obsolete.[295] The fastest passenger rail train built in America was the New York Central "Black Beetle," powered by jet engines mounted on the top of an interurban rail car, in 1966. It moved faster than any American train to date, at 183 mph. It couldn't compete with the ubiquity of traffic or the speed of flight, but designers in Japan studied it to develop the first Shinkansen bullet trains. Digital cameras, GPS navigators, and alarm clocks are all languishing as telephones can now perform those tasks better.[296] America built its cities with inhumane and obsolete labor and energy sources, which we would not imagine using again. In each of these cases old ways are being replaced by new ones, not because of lack of the old, but because the new is better.

TECHNOLOGICAL AND POLICY REVOLUTIONS DON'T PREVAIL BY CONDEMNING THE status quo. They prevail by giving us what the status quo gave us all along, but in a much better way. Automotive traffic replaced horse-drawn traffic. The internet is replacing both mail and malls. Transportation choice can replace traffic dependence.

In the last few decades, America's transportation policy has begun to acknowledge the existence of transit, walking, and biking. Over half of metro rail transit networks in the US have been built since the 1960s. Transportation funding bills in the 1990s have mentioned walking and biking as funding goals, though many of the projects have been in relation to the existing road network. With the rise and fall of the economy, America has turned to and away from biking, transit, and energy efficient vehicles. We still aren't offering real choice, because we have built for traffic as the only way to get around our regions. We have not built for transportation choice—yet.

In the last decade, many of the counties and regions with installed rail transit networks have adopted transit-oriented zoning policies. Most, if not all, of these policies affect zoning densities within no more than a half mile of stations. There are between four hundred and seven hundred of these policies, and I have not yet studied them in detail. They are an encouraging first step toward nationwide restoration of transportation and land use choice, but this book proposes more. If transit can affect land use for a ten-minute walking trip, why not for a ten- or twenty-minute bike trip?

The choice is ours. Walking, biking, and placemaking for a human-scaled existence can be vigorous tools of creativity and prosperity or can be neglected and treated as irrelevant. We can find creativity, prosperity, and fulfillment only at traffic origins and destinations, or we can find it everywhere. It depends on how much choice we offer in our transportation and landscape.

**NEARLY ALL AMERICANS HAVE GROWN UP IN PLACES WHERE TRAFFIC WAS THE ONLY** way to get around. Walkability, bikeability, transit-oriented development, and transit are not for everyone, but the way in which traffic is given all the perks of law and land use—without uniformly better performance—shows that America has left value on the table. This book shows how we can pick up that value; revitalize our cities, communities, and suburbs; and grow our economy anew.

# Notes

1. An homage to Henry Kissinger's turn of phrase regarding Vladimir Putin and the Ukraine Crisis of 2014. See Kissinger, "How the Ukraine Crisis Ends," *Washington Post*, March 5, 2014, https://www.washingtonpost.com/opinions/henry-kissinger-to-settle-the-ukraine-crisis-start-at-the-end/2014/03/05/46dad868-a496-11e3-8466-d34c451760b9_story.html.

2. "Congestion: BTS TranStats Table 1–69: Annual Person-Hours of Highway Traffic Delay Per Auto Commuter," Bureau of Transportation Statistics (n.d.), *National Transportation Statistics*, https://www.bts.gov/content/annual-person-hours-highway-traffic-delay-person; VMT BTS Transtats Table 1–35, US Vehicle Miles (n.d.), https://www.bts.gov/archive/publications/national_transportation_statistics/table_01_35; Employment: BLS Household Data Annual Averages, "Employment Status of the Civilian Noninstitutional Population, 1942 to date," accessed September 19, 2025, https://www.bls.gov/cps/aa2007/cpsaat1.pdf.

3. Thomas C. Prang, "The African Ape-Like Foot of *Ardipithecus Ramidus* and Its Implications for the Origin of Bipedalism," *eLife 8* (April 2019): e44433.

4. Lisa Richmond, "The Composition, Publication, and Influence of Gilberte Périer's *La Vie de Monsieur Pascal*" (PhD diss., University of British Columbia, 1998), UBC Theses and Dissertations, https://dx.doi.org/10.14288/1.0099245.

5. Serge Bellu, *Histoire mondiale de l'automobile* (Flammarion-Pere Castor, 1998), 10.

6. Tony Hadland and Hans-Erhard Lessing, *Bicycle Design: An Illustrated History* (MIT Press, 2016).

7. John Roney and Robert Hard, "The Beginnings of Maize Agriculture," *Archaeology Southwest* 23, no. 1 (Winter 2009): 4; Andrew F. Smith, *The Tomato in America: Early History, Culture, and Cookery* (University of South Carolina Press, 1994); Michael Nee, "The Domestication of Cucurbita (*Cucurbitaceae*)," *Economic Botany* 44, no. 3 (July 1990): 56–68.

8. Rebecca Taft, "The Trading Path and North Carolina," *Journal of Backcountry Studies* 3, no. 2 (2010): https://libjournal.uncg.edu/index.php/jbc/article/viewFile/26/15.

9. Andrew F. Burghardt, "The Origin and Development of the Road Network of the

Niagara Peninsula, Ontario, 1770–1851," *Annals of the Association of American Geographers* 59, no. 3 (1969): 417–40.

10. Eric Jaffe, *The King's Best Highway: The Lost Story of the Boston Post Road, the Route that Made America* (Scribner, 2013).

11. Jon Milan, *Old Chicago Road: US-12 from Detroit to Chicago* (Arcadia Publishing, 2011).

12. Jaffe, *King's Best Highway*.

13. National and Language Boundary Map via William C. Sturtevant, "Early Indian Tribes, Culture Areas, and Linguistic Stocks," University of Texas Libraries, http://www.lib.utexas.edu/maps/national_atlas_1970/ca000097.jpg; trails data via Fred Ramen, *A Historical Atlas of North America Before Columbus* (Rosen Young Adult, 2004); Paul A. W. Wallace, *Indian Paths of Pennsylvania* (Commonwealth of Pennsylvania, Pennsylvania Historical and Museum Commission, 1993); William C. Mills, *Archeological Atlas of Ohio, 1914* (Ohio State Archeological and Historical Society, 1914*)*, see *Indian Trails and Towns in Ohio*, http://www.railsandtrails.com/Maps/OhioArch1914/trails .htm; Charles M. Davis, *Readings in the Geography of Michigan* (Ann Arbor Publishers, 1964); D. E. Shapiro, "Indian Tribes and Trails of the Chicago Region: A Preliminary Study of the Influence of the Indian on Early White Settlement" (PhD diss., University of Chicago, 1929).

14. "The Black Death: Bubonic Plague," The Middle Ages.Net, http://www.themid dleages.net/plague.html.

15. Lauro Martines, *The Furies: War in Europe, 1450–1700* (Bloomsbury Press, 2013).

16. Mortality estimates based on Mexico City after contacts with the Spanish via Brian M. Fagan, *Kingdoms of Gold, Kingdoms of Jade: The Americas Before Columbus* (Thames & Hudson, 1991).

17. The first reported incident of syphilis in Europe was during the French siege of Naples in 1494, indicating something about the cosmopolitan attractions of European soldiers and sailors, and their sexual mores. David Farhi and Nicolas Dupin, "Origins of Syphilis and Management in the Immunocompetent Patient: Facts and Controversies," *Clinics in Dermatology* 28, no. 5 (Sep.–Oct. 2010): 533–38.

18. Russell Thornton, *American Indian Holocaust and Survival: A Population History Since 1492* (University of Oklahoma Press, 1990), 26–32; Campbell Gibson, "Population of the 100 Largest Cities and Other Urban Places in the

United States: 1790 to 1990," Census Working Paper Number POP-WP027, June 1998; Access Genealogy, "1847 Indian Population of the United States and Territory," accessed September 19, 2025, http://www.accessgenealogy.com/na tive/races/indian_population_united_states_territory.htm; Access Genealogy, "Indian Census of 1853–1890," accessed September 18, 2025, http://www .accessgenealogy.com/native/census/condition/indian_census_1853_1890 .htm#1890; InfoPlease, "Colonial Population Estimates," updated August 5, 2020, http://www.infoplease.com/ipa/A0004979.html; United States Census Bureau, "Colonial and Pre-Federal Statistics," accessed July 9, 2025, http:// www2.census.gov/prod2/statcomp/documents/CT1970p2-13.pdf; Campbell Gibson and Kay Jung, "Historical Census Statistics on Population Totals By Race, 1790 to 1990, and By Hispanic Origin, 1970 to 1990, for the United States, Regions, Divisions, and States. Population Division," *Working Paper Series* no. 56 (2002); Rand McNally and Company, "U.S. Population 1780–1890; By state, 1890," David Rumsey Map Collection, http://www.davidrumsey.com /luna/servlet/detail/RUMSEY~8~1~20750~560031:Population-of-the-United -States-at-.

19. Rebecca Solnit, *Wanderlust: A History of Walking* (Verso, 2002), 233.

20. Cheryl Breen, Jekabs Vittands, and Daniel O'Brien, "The Boston Harbor Project: History and Planning," *Civil Engineering Practice* 9, no. 1 (1994): 11–32.

21. Leonard Metcalf and Harrison Eddy, *American Sewerage Practice: Vol. I: Design of Sewers* (McGraw Hill, 1914).

22. For example, fast-growing Chicago reversed the flow of its river in 1900 just to keep its sewage away from its water intakes on Lake Michigan. Per Stanley A. Changnon and Joyce M. Changnon, "History of the Chicago Diversion and Future Implications," *Journal of Great Lakes Research* 22, no. 1 (1996): 100–118. When Philadelphia built its pioneering water pumping plant on the pastoral Schuylkill in 1812, it delivered clean, fresh river water to a thirsty city of one hundred thousand. The plant proved to be a godsend to citizens whose water supply had previously consisted of a variety of stagnant wells in the bottomlands of Center City. By 1860, the Schuylkill had been developed as an industrial powerhouse, and its watershed had been developed as a residential exurb, with all those outhouses. The city's water supply was more toxic than ever, with typhoid and cholera as leading causes of death in the city. Philadelphia needed more than pumping, it needed treatment, and so installed one

of the first sand filtration systems in the US in 1902. Per Adam Levine, "The History of Philadelphia's Watersheds and Sewers," *Philadelphia Water Department Blog*, accessed August 18, 2025, https://classic.waterhistoryphl.org/index .htm. Charleston's 1857 attempt at municipal sewage disposal relied on the tides to clear out its collector sewer, built perfectly flat for three miles under the city between the Cooper and Ashley rivers. This worked if the tidal bore was strong enough to flush out the pipe. But it was not. The sewage of forty thousand Charlestonians soon backed up into the streets once the collector filled up. Christina R. Butler, *Lowcountry at High Tide: A History of Flooding, Drainage, and Reclamation in Charleston, South Carolina* (University of South Carolina Press, 2020).

23. Jim Ulvog, "10 Leading Causes of Death in 1850 and 2000," Nonprofit Update, October 21, 2010, http://nonprofitupdate.info/2010/10/21/10-leading-causes-of -death-in-1850-and-2000-2/. For the 1850, 1900, and 2000 leading causes of death: Census Bureau, "Leading Causes of Death 1900–1998," accessed August 18, 2025, https://archive.cdc.gov/www_cdc_gov/nchs/data/dvs/lead1900_98 .pdf. For the 1950 leading causes of death and for the death rates, see "The Stork and the Grim Reaper USA," accessed August 18, 2025, archived at https:// web.archive.org/web/20120904062455/http://www.populationeducation.org /docs/300millionlessons/sgrusa.pdf.

24. The "Why Did the Chicken Cross the Road?" joke was written in the 1850s and was then considered hilarious. See *The Knickerbocker, or The New York Monthly*, March 1847, 283.

25. Richard Jones, ed., *Manure Matters: Historical, Archaeological and Ethnographic Perspectives* (Ashgate, 2012).

26. Eric Morris, "From Horse Power to Horsepower," *Access Magazine* 1, no. 30, 2–10.

27. Trip Williams, "Tanning Leather," Blog by user Dragoona, n.d., http://www .alpharubicon.com/primitive/tanningdragoona.htm.

28. Alonzo Lewis, *History of Lynn, Essex County, Massachusetts, Including Lynnfield, Saugus, Swampscott, and Nahant, 1629–1864*, vol. 1 (G. C. Herbert, 1890).

29. Garrick E. Louis, "A Historical Context of Municipal Solid Waste Management in the United States," *Waste Management & Research* 22, no. 4 (August 2004): 306–22.

30. Martin V. Melosi, "National Historic Landmarks nomination, Fresno Sanitary Landfill (1937)," August 7, 2001, https://www.fresno.gov/wp-content

/uploads/2023/06/2001-National-Historic-Landmark-Nomination-Fresno
-Sanitary-Landfill.pdf.

31. Enrique Alonso and Ana Recarte, "Pigs in New York City; a Study on 19th
Century Urban 'Sanitation,'" *Case Study for the Friends of Thoreau Environmental Program, Research Institute of American Studies, University of Alcala,
Spain*, accessed September 18, 2025, https://institutofranklin.net/en/research
/research-groups/friends-thoreau.

32. Christopher Leinberger, *The Option of Urbanism: Investing in a New American
Dream* (Island Press, 2009).

33. Hal Marcovitz, *The Declaration of Independence: Forming a New Nation* (Simon
and Schuster, 2014).

34. Thomas Jefferson, *The Autobiography of Thomas Jefferson* (Good Press, 2023).

35. Michael Passanisi, "Development of the Boston Area Highway System," *Historical Journal of Massachusetts* 23, no. 2 (Summer 1995): 166.

36. Christopher Browne, *A New Mapp of New England and Annapolis and the Coun-
trys [sic] Adjacent* (London, 1690), Norman B. Leventhal Map & Education
Center, accessed September 18, 2025, https://collections.leventhalmap.org
/search/commonwealth:3f462t03c.

37. Allan Wolf, *Zane's Trace* (Candlewick Press, 2007), includes information about
the development of clearing standards.

38. Maggie Stewart-Zimmerman, "Historic American Highways," last updated July
7, 2009, http://homepages.rootsweb.ancestry.com/~maggieoh/highway.html.

39. Alice Morse Earle, *Stage-Coach and Tavern Days* (DigiCat, 2022).

40. The first "public" transportation in any city was not really public. Blaise Pascal's
system of carriages offered secure rides through the streets of Paris to nobil-
ity and gentry for the low price of five sous (one quarter of a livre, or $1.00 as
measured in 2010 dollars). Of course, France had a few governments between
now and then, but we can use the stability and duration of gold price records
to convert between livres then and dollars today. I could be off by an order of
magnitude or more. This proto-transit served the wealthiest fifth of the population of the French capital between 1663 and 1675. See Leon Bernard, "Technological Innovation in Seventeenth-Century Paris," in *The Pre-Industrial Cities & Technology Reader*, ed. Colin Chant (Routledge, 1998), 157.

41. David F. D'Alessandro, Paul D. Romary, Lisa J. Scannell, and Bryan Woliner,
*MBTA Review* (November 1, 2009): 8–12, accessed September 18, 2025, https://

ti.org/pdfs/MBTA_Review_2009.pdf.

42. The first stagecoach in Washington, DC, opened in 1804, to link established Georgetown with the growing capital. The stagecoach used much the same route as today's Circulator run by Metro in the 2000s. Ray Kukulski and Bill Gallagher, "Washington's Trolley System: The Forces that Shaped It, The Benefits that Were Created and The Elements that Caused Its Demise," *Presentation for National Parks Service*, March 1, 2009.

43. This is lower than Manhattan's current density of 110 people per acre, and almost half of the peak density of 160 people per acre in 1910. The safety elevator was invented in 1852, so no building was over five stories tall in 1820. The first skyscraper in New York, The Equitable Life Building, was built in 1870 and was seven stories tall. Equitable Life Assurance Society of the United States, "The Elevator Did It," *The Equitable News: An Agents' Journal* 23 (November 1901): 11.

44. Randel Farm Map of 1820, the first map showing the speculative grid of blocks that occupied the rest of Manhattan. See William Bryk, "The Treasure Map of Manhattan," *Last Exit Blog*, last updated December 12, 2007, accessed August 18, 2025, at https://web.archive.org/web/20080224112259/http://lastexit mag.com:80/article/the-treasure-map-of-manhattan.

45. Jeremy Atack and Robert A. Margo, "'Location, Location, Location!' The Price Gradient for Vacant Urban Land: New York, 1835 to 1900," *The Journal of Real Estate Finance and Economics* 16, no. 2 (March 1998): 151–72.

46. New York Department of Parks and Recreation, "Washington Square Park," accessed July 14, 2025, http://www.nycgovparks.org/parks/washington-square -park/history.

47. The first omnibuses were horse-drawn wagons carrying multiple passengers along fixed routes with fixed schedules, and were introduced in Manchester (1824), Berlin (1825), and Nantes (1826), extending from horse-drawn taxicab services. See Fabienne Maniére, "10 Août 1826: Premiers omnibus à Nantes," *Le Média de L'Histoire*, July 5, 2023, https://www.herodote .net/10_aout_1826-evenement-18260810.php. One stop for the transit service to the baths at Nantes stopped next to a hatter named M. Omnes. His shop's sign read "OMNES OMNIBUS," Latin for "Everything for Everyone." The operator of the transit line, Stanislaus Baudry, liked the name and took it for his transit line. Maxwell Gordon, *A History of the World's Roads and of the Vehicles that Used Them* (Bedford/St. Martin's, 1992), 129.

48. Carl Holliday, "Street Cars Eighty-Five Years Old," *Transit Journal* 50, no. 20 (Jan–June 1920): 904.

49. Thomas Ryder, "The Herdic Coach," *The Carriage Journal* 25, no. 4, (Spring 1988): 171–74.

50. George E. Waring, Jr., compiler, *1880 Census Report Volume 18 & 19 Reports on the Social Statistics of Cities* (Arno Press, 1970).

51. Lay, *Ways of the World*, 129.

52. Sherry Wilding-White, "The Abbot-Downing Company and the Concord Coach," *Industrial Archaeology* 20, no. 1 (1994): 2.

53. William Sloane Kennedy, *Wonders and Curiosities of the Railway; or, Stories of the Locomotive in Every Land* (Chicago, 1884), 169–70.

54. Horsecars had a platform at the front for a driver and single horse or mule and a hitch on either end. Passengers paid for their rides by rolling a nickel (the nickel in 1860 is the equivalent of $1.35 in 2010) to a cashbox along metal rails along either side of the horsecar. See Frank Rowsome and Stephen Maguire, *Trolley Car Treasury: A Century of American Streetcars: Horsecars, Cable Cars, Interurbans, and Trolleys* (McGraw Hill, 1956), 25.

55. "Chronology of Train Control Development," Appendix E, Bureau of Transportation Statistics (n.d.).

56. John R. Prentice, "Horse Trams on Rail and Road," accessed July 14, 2025, https://www.tramwayinfo.com/tramways/Articles/Ehorse.htm.

57. Charles E. Lee, "Sources of Bus History," *The Journal of Transport History* 3 (1956): 152–57.

58. "Horse Car in Washington, D.C.," Prints and Photographs Online Catalog (PPOC), Library of Congress (1893), Reproduction Number: LC-USZ62-16467 (b&w film), http://www.loc.gov/pictures/item/2012646772/.

59. Morris, "From Horse Power to Horsepower."

60. Marion S. Lane and Stephen L. Zawistowski, *Heritage of Care: The American Society for the Prevention of Cruelty to Animals* (Bloomsbury Publishing, 2007).

61. Lay, *Ways of the World*, 132.

62. Morris, "From Horse Power to Horsepower."

63. Morris, "From Horse Power to Horsepower."

64. Jeffrey Michael Flanagan, "On the Backs of Horses: The Great Epizootic of 1872" (PhD diss., William and Mary, 2011), W&M ScholarWorks, https://scholarworks.wm.edu/handle/internal/8403.

65. Flanagan, "On the Backs of Horses."

66. Greg Sabin, "Economy Tanked Over Cows, Horses and Whales," *CNN*, February 2, 2009, http://www.cnn.com/2009/LIVING/wayoflife/02/02/mf.economic .crisis.history/index.html.

67. Flanagan, "On the Backs of Horses."

68. Flanagan, "On the Backs of Horses."

69. Rowsome and Maguire, *Trolley Car Treasury*.

70. William Middleton, *The Time of the Trolley* (Kalmbach Publishing, 1967), 78–79; Frederic Nicholas, *McGraw Electric Railway Manual* (McGraw Hill, 1920); *McGraw Electric Railway Directory* (McGraw Hill Publishing Company Inc., 1920); George W. Hilton and John F. Due, *The Electric Interurban Railways in America* (Stanford University Press, 1960); Robert W. Mann, *The Streetcars of Florida's First Coast* (Arcadia Publishing, 2014), 180; Frankfort Roundabout, "Electric Cars," March 31, 1894, 4 (Maysville, KY); *Evening Bulletin*, January 17, 1891, 2; "Street Cars Are Discontinued as Bus Trips Begin," *Owensboro Messenger*, April 15, 1934, 1; Sioux Falls City Council, *Dakota Territory & Sioux Falls Railroads*, presented by Bruce Danielson, https://www.southdacola.com/blog /wp-content/uploads/2015/08/Sioux-Falls-SD-Railroad-history-compilations .pdf; Andrew D. Young, *Veteran & Vintage Transit* (Archway Publishing, 1997), 76; George W. Hilton, *The Cable Car in America* (Stanford University Press, 1997).

71. Middleton, *The Time of the Trolley*, 78–79; Nicholas, *McGraw Electric Railway Manual*; *McGraw Electric Railway Directory*; Hilton and Due, *Electric Interurban Railways in America*; Mann, *Streetcars of Florida's First Coast*; Roundabout, "Electric Cars," 4; *Evening Bulletin*, 2; "Street Cars Are Discontinued," *Owensboro Messenger*, 1; Sioux Falls City Council, *Dakota Territory*; Young, *Veteran & Vintage Transit*; Hilton, *Cable Car in America*; *Tramways & Urban Transit* (September 2000): 348–49.

72. Rich Sampson, "The Return of New Orleans' Transit Legacy," *Community Transportation* 26, no. 3 (October 2018): 6–12.

73. Sampson, "The Return of New Orleans' Transit Legacy."

74. Middleton, *The Time of the Trolley*, 78–79; Nicholas, *McGraw Electric Railway Manual*; *McGraw Electric Railway Directory*; Hilton and Due, *Electric Interurban Railways in America*; Mann, *Streetcars of Florida's First Coast*; Roundabout, "Electric Cars," 4; *Evening Bulletin*, 2; "Street Cars Are Discontinued as

Bus Trips Begin," *Owensboro Messenger*, 1; Sioux Falls City Council, *Dakota Territory*; Young, *Veteran & Vintage Transit*; Hilton, *Cable Car in America*; *Tramways & Urban Transit*, 348–49.

75. Rowsome and Maguire, *Trolley Car Treasury*, 35.

76. Oscar Terry Crosby and Louis Bell, *The Electric Railway in Theory and Practice* (New York, 1893).

77. The story that the first gripman quit after seeing the steep hill he had to pilot, leaving Hallidae to drive the car himself, is probably apocryphal. Keli Dailey, "Cable Car History," San Francisco Municipal Transportation Agency, accessed July 14, 2025, http://sfmta.com/about-sfmta/our-history-and-fleet /sfmta-fleet/cable-cars.

78. Rowsome and Maguire, *Trolley Car Treasury*.

79. Joe Thomson, "Cable Car Systems of the World," *Cable Car Guy Blog*, https:// www.cable-car-guy.com/html/ccarch.html.

80. Rowsome and Maguire, *Trolley Car Treasury*.

81. Crosby and Bell, *The Electric Railway in Theory and Practice*.

82. Z. Andrew Farkas, "Urban Transportation Policy: The Baltimore Experience," in *Handbook of Transportation Policy and Administration*, ed. Jeremy F. Plant, Van R. Johnston, and Cristina E. Ciocirlan (Routledge, 2007).

83. Michael Bobbins, "The Early Years of Electric Traction: Invention, Development, Exploitation," *The Journal of Transport History* 21, no. 1 (2000): 92–101.

84. Casey Cater, *Regenerating Dixie: Electric Energy and the Modern South* (University of Pittsburgh Press, 2019).

85. Joe Thompson, collector, "Selected articles from *Manufacturer and Builder*, 1885–1889," retrieved from *Cable Car Guy Blog*, http://www.cable-car-guy .com/html/ccmanbl3_b.html.

86. Rowsome and Maguire, *Trolley Car Treasury*.

87. Crosby and Bell, *The Electric Railway in Theory and Practice*.

88. Middleton, *The Time of the Trolley*, 78–79; Nicholas, *McGraw Electric Railway Manual*; *McGraw Electric Railway Directory*; Hilton and Due, *Electric Interurban Railways in America*; Mann, *Streetcars of Florida's First Coast*; Roundabout, "Electric Cars," 4; *Evening Bulletin*, 2; "Street Cars Are Discontinued," *Owensboro Messenger*, 1; Sioux Falls City Council, *Dakota Territory*; Young, *Veteran & Vintage Transit*; Hilton, *Cable Car in America*; *Tramways & Urban Transit* (September 2000): 348–49.

89. Llewellyn Park Neighborhood Association, "History," accessed July 14, 2025, http://www.llewellynpark.com/page/13266~93841/History.

90. Steve Lech, *Riverside, 1870–1940* (Arcadia Publishing, 2007).

91. Ted Bazemore, "Inman Park," *The New Georgia Encyclopedia*, last updated August 13, 2018, https://www.georgiaencyclopedia.org/articles/counties-cities -neighborhoods/inman-park/.

92. Robert Benedetto, Jane Donovan, and Kathleen DuVall, *Historical Dictionary of Washington* (Scarecrow Press, 2003), 53.

93. Marian J. Morton, *Cleveland Heights: The Making of an Urban Suburb* (Arcadia Publishing, 2002).

94. Mount Arlington Environmental Resources Inventory Township of Maplewood Essex County, New Jersey (New Jersey Environmental Resources Inventory, 2005).

95. Douglas P. Munro, *Greater Roland Park* (Arcadia Publishing, 2015).

96. Linda K. Williams and Paul S. George, "South Florida: A Brief History," Historical Museum of Southern Florida, accessed September 19, 2025, https:// web.archive.org/web/20100429002717/http://www.hmsf.org/history/south -florida-brief-history.htm.

97. Stephen Kieffer, *Transit and the Twins* (Twin City Rapid Transit Company, 1958), 20.

98. Brad Huff, *The Rise of Portland's Public Transportation System* (2006).

99. William Middleton, *The Interurban Era* (Kalmbach, 1961).

100. Middleton, *The Interurban Era*.

101. William E. Thoms, "Unleashing the Greyhounds: The Bus Regulatory Reform Act of 1982," *Campbell Law Review* 6, no. 1 (Spring 1984): 75.

102. G. J. Christiano, Mark S. Feinman, David Pirmann, and Michael Calcagno, "The 9th Avenue Elevated-Polo Grounds Shuttle," *NYC Subway Blog*, accessed July 14, 2025, http://www.nycsubway.org/wiki/The_9th_Avenue_Elevated -Polo_Grounds_Shuttle.

103. Young Ewing Allison, *The City of Louisville and a Glimpse of Kentucky* (Louisville Committee on Industrial and Commercial Improvement of the Louisville Board of Trade, 1887).

104. Greg Borzo, The Chicago "L" (Arcadia Publishing, 2007).

105. American Public Transport Association, "Public Transportation Ridership Report Fourth Quarter 2023" (American Public Transportation Association, 2024),

accessed September 18, 2025, https://www.apta.com/wp-content/uploads /2023-Q4-Ridership-APTA-Update-1.pdf.

106. Robert Steven Diamond and Brian Garret Kassel, "The Confluence of Four Events that Led to the Creation of the Atlantic Avenue Tunnel: The World's First Subway," brooklynrail.net, accessed September 18, 2025, https://brooklynrail .net/images/aa_tunnel/new_research/oct_09/events_leading_to_tunnel_cre- ation.pdf.

107. John Day and John Reed, *The Story of London's Underground* (Capital Trans- port, 1963).

108. Robert A. Cohen, "The Pneumatic Mail Tubes: New York's Hidden High- way and Its Development," in *Proceedings of the 1st International Sympo- sium on Underground Freight Transport by Capsule Pipelines and Other Tube/ Tunnel Systems*, USPS (August 1999): https://about.usps.com/who-we-are /postal-history/pneumatic-tubes.pdf.

109. Albert Stetson, *The Practicability of Electric Conduit Railways* (Transactions of the American Institute of Electrical Engineers, 1893), 10:627–64.

110. Streetcar companies owned and operated over five hundred amusement parks and resorts at the end of their lines by 1910. See Robert Post, *Urban Mass Tran- sit: The Life Story of a Technology* (Johns Hopkins University Press, 2010).

111. Maury Klein, *The Life and Legend of Jay Gould* (Johns Hopkins University Press, 1997).

112. T. J. Stiles, *The First Tycoon: The Epic Life of Cornelius Vanderbilt* (Knopf, 2009).

113. Maury Klein, *The Life and Legend of E. H. Harriman* (University of North Car- olina Press, 2000).

114. Homer Charles Harlan, *Charles Tyson Yerkes and the Chicago Transportation System* (University of Chicago Press, 1975).

115. William B. Friedricks, *Henry E. Huntington and the Creation of Southern Cali- fornia* (Ohio State University Press, 1992).

116. Rowsome and Maguire, *Trolley Car Treasury*, 141.

117. Details on Widener's Transit Empire in Philadelphia can be found in Charles W. Cheape, *Moving the Masses: Urban Public Transit in New York, Boston, and Philadelphia, 1880–1912*, vol. 31 (Harvard University Press, 1980).

118. John Franch, *Robber Baron: The Life of Charles Tyson Yerkes* (University of Il- linois Press, 2006).

119. George Drury, *The Historical Guide to North American Railroads: Histories,*

*Figures, and Features of More Than 160 Railroads Abandoned or Merged Since 1930* (Kalmbach Publishing, 1994).

120. The "Toonerville Trolley" comics printed between 1908 and 1955 were emblematic of the beloved but dilapidated state of many of the nation's trolleys. See Mildred M. Walmsley, "The Bygone Electric Interurban Railway System," *The Professional Geographer* 17, no. 3 (1965): 1–6.

121. National City Lines, formed in 1936 by car, tire, and oil companies, bought up many trolley networks from their private and public agencies and converted them to bus fleets. This destruction and recreation of America's transit was not a conspiracy so much as a public service. The trolley fleets were dilapidated from fares set too low for too long. Transit agencies did not see themselves as being in the streetcar business so much as the transit business. Riders had little sentimental attachment to the particular vehicles, just a pragmatic need to move passengers around at minimum cost for maximum service. Buses were able to dodge congested lanes while trolleys were bound to their tracks. Buses did not require catenary wires, rails, or substations to operate, so transit agencies no longer had to pay to maintain them. They could use the same parking layover for their buses as their trolleys, retrain their drivers, and get out of the power generation business entirely. Buses were piloted much like cars in the same streets with traffic, and they replaced trolleys in most cities by 1940 as the lower cost alternative. See David Jones, *Mass Motorization and Mass Transit: An American History and Policy Analysis* (Indiana University Press, 2010).

122. See *American Public Transport Association Fact Book 2020*, accessed September 18, 2025, https://www.apta.com/wp-content/uploads/APTA-2020-Fact-Book.pdf.

123. "The U.S. dollar has lost 97% its value since 1774," officialdata.org, updated June 11, 2025, https://www.officialdata.org/us/inflation/1774.

124. Jones, *Mass Motorization and Mass Transit*.

125. Stephen Goddard, *Getting There: The Epic Struggle Between Road and Rail in the American Century* (University of Chicago Press, 1996).

126. Jones, *Mass Motorization and Mass Transit*.

127. Streetcar history: Middleton, *The Time of the Trolley*, 78–79; Nicholas, *McGraw Electric Railway Manual*; *McGraw Electric Railway Directory*; Hilton and Due, *Electric Interurban Railways in America*; Mann, *Streetcars of Florida's First Coast*; Roundabout, "Electric Cars," 4; *Evening Bulletin*, 2; "Street Cars Are

Discontinued," *Owensboro Messenger*, 1; Sioux Falls City Council, *Dakota Territory*; Young, *Veteran & Vintage Transit*; Hilton, *Cable Car in America*; *Tramways & Urban Transit* (September 2000): 348–49; Light Rail history: Michael Taplin, "A world of trams and urban transit," *Light Rail and Modern Tramway* 60, no. 718, supplement (1997): 1–8; American Public Transport Association, "Public Transportation Ridership Report, Fourth Quarter 2018," April 12, 2019, https://www.apta.com/wp-content/uploads/2018-Q4-Ridership-APTA .pdf; Michael Taplin, "The History of Tramways and Evolution of Light Rail," *Light Rail Transit Association*, accessed August 25, 2016, https://lrta.info /archive/mrthistory.html; Leroy W. Demery Jr., "U.S. Urban Rail Transit Lines Opened from 1980: Appendix," accessed November 3, 2013, https:// web.archive.org/web/20131103235042/http://www.publictransit.us/ptlibrary /NorthAmericaRailTransitOpenings/Railopenings_ZAppend_2010.htm; Interurban history: Middleton, *Interurban Era*; Hilton and Due, *Electric Interurban Railways in America*; Heavy Rail history: American Public Transport Association, "Public Transportation Ridership Report Fourth Quarter 2023," accessed September 18, 2025, https://www.apta.com/wp-content/uploads/2023 -Q4-Ridership-APTA-Update-1.pdf; Bus Transit history: *The National Transit Database (NTD)* (PDF), "Title" (August 21, 2015). Archived from the original (PDF), accessed March 4, 2016.

128. Andrew Livesey, *Bicycle Engineering and Technology* (Routledge, 2020).

129. Blood Sweat & Gears C.I.C., "Mud Sweat and Gears Blog: A Pictorial History of the Evolution of The Bicycle," accessed January 12, 2016, http://www.mud sweatngears.co.uk/page_2473200.html.

130. National Museum of Natural History Behring Center, "America on the Move," *General Motors Hall of Transportation*, Smithsonian Institute, accessed July 14, 2025, http://amhistory.si.edu/onthemove/themes/story_69_2.html.

131. The *"Phantom," The Velocipede of the Period* (London, 1899).

132. Albert Augustus Pope bought a Starley Penny-Farthing in England and brought it back to the US for reverse engineering. See Jaffe, *King's Best Highway*.

133. David V. Herlihy, *Bicycle: The History* (Yale University Press, 2004).

134. Blood Sweat & Gears C.I.C., "Mud Sweat and Gears Blog: Pictorial History."

135. Herlihy, *Bicycle: The History*, 184–92.

136. Carlton Reid, *Roads Were not Built for Cars: How Cyclists Were the First to Push for Good Roads & Became the Pioneers of Motoring* (Island Press, 2015).

137. T. D. Denham, "California's Great Cycle-Way," *Good Roads Magazine*, November 1901.

138. Christopher Wells, "The Changing Nature of Country Roads: Farmers, Reformers, and the Shifting Uses of Rural Space, 1880–1905," *Agricultural History* 80, no. 2 (Spring 2006): 143–66.

139. Jaffe, *King's Best Highway*.

140. Richard F. Weingroff, "A Peaceful Campaign of Progress and Reform: The Federal Highway Administration at 100," *Federal Highway Authority Public Roads* 57, no. 2 (Autumn 1993).

141. Max G. Lay, *Ways of the World*.

142. Trichur S. Suryanarayanan and João Lúcio Azevedo, "From Forest to Plantation: A Brief History of the Rubber Tree," *Indian Journal of History of Science* 58, no. 1 (2023): 74–78.

143. Kristin Fetzer, Elizabeth Harvey, Ira Kauderer, and Laura Spina, "Historic Street Paving Thematic District," Philadelphia Register of Historic Places, Philadelphia Historical Commission, accessed September 18, 2025, https://www .phila.gov/media/20190213131359/Thematic-District-Street-Paving.pdf.

144. An early 1700s paving agreement in Philadelphia held that property owners were to pave the roads next to their houses in stones to the centerline. See Russell Frank Weigley, Nicholas B. Wainwright, and Edwin Wolf, eds., *Philadelphia: A 300-Year History* (W. W. Norton & Company, 1982).

145. Light Rail Now! Publication Team, "Light Rail Now! MythBusters Weblog," *Light Rail Now! Light Rail Progress*, accessed July 14, 2025, http://www.lightrail now.org/myths/m_mythlog001.htm.

146. This was the discovery and innovation of John MacAdam and his competitor Thomas Telford, who saw the destructive forces of millions of tiny insults on pavers and proposed that good pavements were made of pebbles smaller than the width of the wheels, not larger. If the wheel applies force evenly across the width of pebbles in the aggregate, they are less liable to fracture than large stones. These roads were smoother, cheaper, and more durable than stone or cobble roads, which were well suited to the recent revolution in self-propelled vehicles. See Lay, *Ways of the World*.

147. Clay McShane, *Down the Asphalt Path: The Automobile and the American City* (Columbia University Press, 1995).

148. Gordon K. Ray, "History and Development of Concrete Pavement Design,"

*Journal of the Highway Division* 90, no. 1 (January 1964): 79–101.

149. Peter Norton, *Fighting Traffic: The Dawn of the Motor Age in the American City* (MIT Press, 2008), 200.

150. David Mozer, "Chronology of the Growth of Bicycling and the Development of Bicycle Technology," in "Bicycle History (& Human Powered Vehicle History)," International Bicycle Fund blog, accessed September 18, 2025, http://www.ibike.org/library/history-timeline.htm.

151. David Rubinstein, "Cycling in the 1890s," *Victorian Studies* 21, no. 1 (Autumn 1977): 47–71.

152. Eduardo Porter, *The Price of Everything: Solving the Mystery of Why We Pay What We Do* (Portfolio Press, 2011).

153. The Model T Ford Club International, Inc., "Original Model T Prices by Model and Year," updated 2000, accessed March 24, 2016, archived link at https://web.archive.org/web/20160324182622/https://modelt.org/index.php?option=com_content&view=article&id=11:original-model-t-ford-prices-by-model-and-year&catid=5:history-and-lore&Itemid=1.

154. Bureau of Labor Statistics, "Labor and Household characteristics between 1934 and 1936," https://www.bls.gov/.

155. Jaffe, *King's Best Highway.*

156. "Hog River Journal: Exploring CT History," *The Horseless Era Arrives*, accessed July 14, 2025, http://www.kcstudio.com/colha98.html.

157. Jacob R. Mecklenborg, *Cincinnati's Incomplete Subway: The Complete History* (Arcadia Publishing, 2010).

158. American Public Transport Association, "Public Transportation Ridership Report Fourth Quarter 2023."

159. Beverly Rae Kimes and Henry Austin Clark Jr., *The Standard Catalogue of American Cars, 1805–1942*, 2nd ed. (Krause Publications, 1985), 853; Gary Levine, *The Car Solution: The American Steam Car Comes of Age* (Horizon Press, 1974).

160. Dorothy S. Brady, "Relative Prices in the Nineteenth Century," *The Journal of Economic History* 24, no. 2 (June 1964): 145–203.

161. McShane, *Down the Asphalt Path.*

162. Karel Williams, Colin Haslam, Sukhdev Johal, and John Williams, *Cars: Analysis, History, Cases* (Berghahn Books, 1994).

163. The Jitney boom occurred right after a recession and was received by transit and taxi providers of the day every bit as fondly as Uber or Lyft is by taxi operators

or transit agencies are today. See Adam Hodges, "'Roping the Wild Jitney': The Jitney Bus Craze and the Rise of Urban Autobus Systems," *Planning Perspectives* 21, no. 3 (July 2006): 253–76.

164. Jones, *Mass Motorization and Mass Transit*.

165. Norton, *Fighting Traffic*.

166. "Cyclist Dies in Collision," *New York Times*, September 3, 1896.

167. Lay, *Ways of the World*, 132.

168. "Highway Statistics Summary to 1995," Federal Highway Administration, accessed July 14, 2025, http://www.fhwa.dot.gov/ohim/summary95/section2 .html.

169. Rowsome and Maguire, *Trolley Car Treasury*, 10.

170. Norton, *Fighting Traffic*.

171. Norton, *Fighting Traffic*, 96.

172. Andres Duany, Elizabeth Plater-Zyberk, and Jeff Speck, *Suburban Nation: The Rise of Sprawl and the Decline of the American Dream* (North Point Press, 2000).

173. Famously, in the words of Charles Wilson, CEO of General Motors, during congressional confirmation hearings to become Secretary of Defense: "I thought what was good for the country was good for General Motors and vice versa." See Historical Office: Office of the Secretary of Defense, "Charles E. Wilson," accessed September 18, 2025, https://history.defense.gov/Multi media/Biographies/Article-View/Article/571268/charles-e-wilson/#:~:text =During%20the%20hearings%2C%20when%20asked,General%20Motors %20and%20vice%20versa.%22.

174. Norton, *Fighting Traffic*.

175. VMT: "FHWA (2024) Motor Vehicle Traffic Fatalities & Fatality Rate: 1899– 2023," accessed September 18, 2025, https://cdan.dot.gov/tsftables/Fatalities %20and%20Fatality%20Rates.pdf; see Federal Highway Administration's government cost of traffic from HF-210 Series 1921–1995 (2018) "Highway Statistics Summary to 1995," last updated June 3, 2025, https://www.fhwa.dot.gov /ohim/summary95/section4.html; and HF-10, Office of Highway Policy Information, "2012—Quick Find, Highway Finance," accessed September 18, 2025, archived at https://www.fhwa.dot.gov/policyinformation/quickfinddata /qffinance.cfm; Population: United States Census Bureau, "Historical Population Change Data, 1910–2020," accessed September 18, 2025, https://www.census

.gov/data/tables/time-series/dec/popchange-data-text.html; GDP: Aaron O'Neill, "Annual Gross Domestic Product and real GDP in the United States from 1929 to 2022," Statista, accessed July 14, 2025, https://www.statista.com /statistics/1031678/gdp-and-real-gdp-united-states-1930-2019/; Number of automobiles: "Highway Statistics Summary to 1995."

176. Peter Zheutlin, "Women on Wheels: The Bicycle and the Women's Movement of the 1890s," *Annie Londonderry: The First Woman to Bicycle Around the World*, accessed July 14, 2025, http://www.annielondonderry.com/women Wheels.html.

177. "Road Miles: Bureau of Transportation Statistics," *National Transportation Statistics*, accessed September 18, 2025, archived at http://www.rita.dot.gov /bts/sites/rita.dot.gov.bts/files/publications/national_transportation_statis- tics/html/table_01_04.html; Population: United States Census Bureau, "Population, Housing Units, Area Measurements, and Density: 1790 to 1990," accessed September 18, 2025, https://www2.census.gov/programs-surveys /decennial/1990/tables/cph-2/table-2.pdf; US Census, "National Intercensal Estimates (2000–2010)," accessed September 19, 2025, archived at https://web .archive.org/web/20150708233701/https://www.census.gov/popest/data/inter censal/national/nat2010.html; Bus Route and Rail Transit Miles: *APTA Transit Data Factbook, Appendix A*, accessed September 19, 2025, https://www.apta .com/wp-content/uploads/Resources/resources/statistics/Documents/Fact Book/2014-APTA-Fact-Book-Appendix-A.pdf; Canal Miles: HomeTown- Locator, Inc, "List of Canals by State," accessed December 1, 2013, https:// www.hometownlocator.com/; US Census Statistical Abstracts 1878–2012, accessed September 19, 2025, https://www.census.gov/library/publications /time-series/statistical_abstracts.html; Railroad Miles: US Census Statistical Abstracts 1878–2012, accessed September 19, 2025, https://www.census.gov /library/publications/time-series/statistical_abstracts.html; National Trans- portation Statistics, accessed September 19, 2025, https://web.archive.org /web/20170628150615/http://www.rita.dot.gov/bts/sites/rita.dot.gov.bts/files /publications/national_transportation_statistics/html/table_01_01.html.

178. Rolf Pendall, Christopher Hayes, Taz George, Zach McDade, Casey Dawkins, SikJeon Jae, Eli Knaap, Evelyn Blumenberg, Gregory Pierce, and Michael Smart, "Driving to Opportunity: Understanding the Links Among Transpor- tation Access, Residential Outcomes, and Economic Opportunity for Housing

Voucher Recipients," *The Urban Institute*, March 2014, http://www.urban .org/UploadedPDF/413078-Driving-to-Opportunity.pdf.

179. Analysis, Inc., "Trucking Industry Facts, 2010," accessed September 18, 2025, https://www.analysis-inc.com/reference/trucking-industry-facts-2010/.

180. Hal Kane, *Triumph of the Mundane: The Unseen Trends that Shape Our Lives and Environment* (Island Press, 2000).

181. Justin Glow, "The World's Messiest Cars," *Gadling Blog*, May 25, 2007, http:// gadling.com/2007/05/25/the-worlds-messiest-cars/.

182. Rodney E. Slater, "The National Highway System: A Commitment to America's Future," *Public Roads* 59, no. 4 (n.d.): 2–6.

183. Earl Swift, *The Big Roads: The Untold Story of the Engineers, Visionaries, and Trailblazers Who Created the American Superhighways* (Houghton Mifflin Harcourt, 2011).

184. Sofie Van den Waeyenberg and Luc Hens, "Crossing the Bridge to Poverty, with Low-Cost Cars," *Journal of Consumer Marketing* 25, no. 7 (October 2008): 439–45.

185. Dayna Evans, "Free Parking is Killing Cities," *Bloomberg*, August 31, 2021, https:// www.bloomberg.com/news/features/2021-08-31/why-free-parking-is-bad -according-to-one-ucla-professor.

186. "Road Traffic Deaths: Data by WHO Region," World Health Organization, February 9, 2021, accessed March 2, 2024, https://apps.who.int/gho/data/view .main-wpro.RoadTrafficDeathREG?lang=en.

187. Specifically, William Phelps Eno, wealthy, self-appointed traffic planner who acted as the first arbiter of traffic rules a year after the first walker was killed by traffic in his native New York. He invented the circular intersection right-of-way pattern. Per Norton, *Fighting Traffic*, and John A. Montgomery, *Eno—The Man and the Foundation: A Chronicle of Transportation* (Eno Center for Transportation, 1988).

188. "Highway Statistics Summary to 1995," section V, "Roadway, Extent, Characteristics and Performance. Motor Vehicle Traffic Fatalities, National Summary" (Table FI-200); "Highway Statistics Summary to 1995"; see "Funding for Highways, All Units of Government (Table HF-210)," in "Highway Statistics Summary to 1995"; American Automobile Association, "Your Driving Costs: How Much Are You Really Paying to Drive? 2010 Edition," accessed September 18, 2025, https://exchange.aaa.com/wp-content/uploads/2012/04/201048935480

.Driving-Costs-2010.pdf; United States Census Bureau, Population Estimates, [n.d], National Intercensal Estimates (2000–2010); GDP: John W. Kendrick, *Productivity Trends in the United States* (Princeton University Press, 1961).

189. Fetzer, Harvey, Kauderer, and Spina, "Historic Street Paving."

190. "Highway Statistics Summary to 1995," section III Highway Finance (Table HF-210).

191. "Highway Statistics Summary to 1995," section V, "Roadway, Extent, Characteristics and Performance: Motor Vehicle Traffic Fatalities, National Summary," (Table FI-200) and US Census Bureau 2012 Statistical Abstract, Law Enforcement, Courts & Prisons, accessed September 18, 2025, https://www.census.gov/library/publications/2011/compendia/statab/131ed/law-enforcement-courts-prisons.html.

192. See census data for 2010 in "TIGER/Line Shapefiles," updated October 9, 2024, https://www.census.gov/geographies/mapping-files/time-series/geo/tiger-line-file.html.

193. Mary R. McCorvie and Christopher L. Lant, "Drainage District Formation and the Loss of Midwestern Wetlands, 1850–1930," *Agricultural History* 67, no. 4 (January 1993): 13–39.

194. VaDOT, "Route 1 in Dumfries, Virginia, Time Series Photostream," accessed July 14, 2025, https://www.flickr.com/photos/vadot/albums/72157632993008741/.

195. Assuming compact car of three thousand pounds and 31 mph, based on "Momentum, Kinetic Energy, and Bullets," by user Scoob, *PhysicsForums Group Blog*, December 9, 2003, http://www.physicsforums.com/showthread.php?t=10591.

196. InfoPlease, "Most Congested Roads," updated August 5, 2020, http://www.infoplease.com/world/transportation/most-congested-roads.html; Thingstock, "Top 10 Most Dangerous Roads," *Destination America Blog*, accessed May 25, 2014, http://destinationamerica.tumblr.com/post/21336189785/top-10-most-dangerous-roads.

197. Yogi Berra, *The Yogi Book: "I Really Didn't Say Everything I Said"* (Workman Publishing, 2010).

198. Fatality crash time data available at Fatality Analysis Reporting System (FARS), accessed July 14, 2025, http://www.nhtsa.gov/FARS. Trip data available via the 2009 National Household Travel Survey, accessed July 14, 2025, http://nhts.ornl.gov/; Florida Highway Safety and Motor Vehicles, "Traffic Crash Statistics

Report 2010," accessed September 18, 2025, https://www.flhsmv.gov/pdf /crashreports/crash_facts_2010.pdf.

199. Mohammad Naim Rastgoo, Bahareh Nakisa, Andry Rakotonirainy, Vinod Chandran, and Dian Tjondronegoro, "A Critical Review of Proactive Detection of Driver Stress Levels Based on Multimodal Measurements," *ACM Computing Surveys (CSUR)* 51, no. 5 (September 2018): 1–35.

200. David Smith, "The Bicycle Driver," *Cranked Magazine* 5, 22–25, accessed July 14, 2025, http://crankedmag.wordpress.com/issues/issue-5/the-bicycle-driver.

201. "Vancouver Cyclist Charged $3,700 for Damage to Car That Hit Him," *Canadian Broadcasting Corporation*, March 31, 2022, https://www.cbc.ca/news /canada/british-columbia/vancouver-bc-cyclist-icbc-insurance-no-fault-1.6 403817.

202. Greg Bluestein, "Wreckage of Easter Crash Hits 3 Atlanta Families," Herald.Net, May 7, 2009, https://www.heraldnet.com/news/wreckage-of-easter -crash-hits-3-atlanta-families/.

203. Sandra Rose, "Survivor Recalls Fatal Easter Crash," *Sandra Rose Blog*, April 27, 2009, http://sandrarose.com/2009/04/survivor-recalls-fatal-easter-crash.

204. "Aimee Michael Sentenced To 36 Years In Prison For Deadly Crash," *WSB-TV Atlanta 2*, November 5, 2010, https://www.wsbtv.com/news/aimee -michael-sentenced-to-36-years-in-prison-for-/241592269/.

205. Ralph Ellis, "Jaywalkers Take Deadly Risks," *Atlanta Journal-Constitution*, July 26, 2011, https://www.ajc.com/news/local/jaywalkers-take-deadly-risks /2BxIEgcQtM4P7WEzCXWEgP/.

206. Thomas Wheatley, "Cobb County Mom Whose Child was Killed While Crossing the Street Gets Convicted of Vehicular Homicide: What Can We Learn From the Story of Raquel Nelson?," *Creative Loafing Atlanta*, July 22, 2011, http://clatl.com/freshloaf/archives/2011/07/22/cobb-county -mom-whose-child-was-killed-while-crossing-the-street-gets-convicted-of -vehicular-homicide.

207. Tanya Snyder, "Georgia Mom Convicted of Vehicular Homicide for Crossing Street with Kids," *Streetsblog USA*, July 14, 2011, https://usa.streetsblog .org/2011/07/14/mother-convicted-of-vehicular-homicide-for-crossing-street -with-children.

208. At the Central Park side of Central Park West Avenue and 74th Street, near the Northeast corner, at 40.7775 latitude, -73.9748 longitude.

209. "Death of Henry Bliss After Stepping Off a Streetcar, Killed by an Electric Taxi," *New York Times*, September 14, 1899, 1.

210. Officer Down Memorial Page, "Patrolman Thomas Meagher," accessed July 14, 2025, http://www.odmp.org/officer/16942-patrolman-thomas-meagher. The nature of their killings depends on the contemporary meaning of the word "car." "Car" may have referred to "streetcar." Baltimore police officer Alonzo Bishop may have been the first American killed in a collision between two cars on August 29, 1899. See "Police Officer Alonzo B. Bishop," Officer Down Memorial Page, accessed September 18, 2025, https://www.odmp.org/officer/1879-police-officer-alonzo-b-bishop.

211. "Ex-Mayor Edson Dead; Began Successful Career as a Country School Teacher," *New York Times*, September 25, 1904, https://www.nytimes.com/1904/09/25/archives/exmayor-edson-dead-began-successful-career-as-a-country-school.htmlquery.nytimes.com/gst/abstract.html?res=9F02E6D8123AE733A25756C2A96F9C946597D6CF.

212. David Zipper, "Crash Course: News Organizations Need to Relearn How to Cover Car Collisions—Especially When the Victims Are on Foot," *Slate*, May 18, 2022, https://slate.com/business/2022/05/media-car-crashes-washington-post-pedestrians.html?fbclid=IwAR3MCAOPJlDVFQH9v6ammEKAtBFgns_jWlgFXVX-5zg_y9r6ucEesg3RKDc.

213. Organization for Economic Co-operation and Development (OECD), "OECD Data / Road Accidents," December 15, 2023, accessed August 8, 2024, https://www.oecd.org/en/data/indicators/road-accidents.html; David Leonhardt, "The Rise in U.S. Traffic Deaths: What's Behind America's Unique Problem with Vehicle Crashes?," *New York Times*, December 11, 2023; Centers for Disease Control, "Accidents or Unintentional Injuries," last updated January 15, 2025, www.cdc.gov/nchs/fastats/accidental-injury.htm; National Highway Traffic Safety Administration, "Motor Vehicle Traffic Fatalities and Fatality Rates, 1899–2020," accessed September 18, 2025, https://cdan.dot.gov/tsftables/Fatalities%20and%20Fatality%20Rates.pdf; Fatal Car Accident Statistics, "Fatality Analysis Reporting System 2010 and 2011," archived from the original on October 26, 2009, https://www-fars.nhtsa.dot.gov/Main/index.aspx; US Census Bureau, "Census Bureau Projects U.S. And World Populations on New Year's Day," December 30, 2019, https://www.census.gov/newsroom/press-releases/2019/new-years-2020.html; NHTSA, "2020 Fatality Data

Show Increased Traffic Fatalities During Pandemic," June 3, 2021, https://www
.nhtsa.gov/press-releases/2020-fatality-data-show-increased-traffic-fatalities
-during-pandemic; "U.S. Population 1950–2025," Macrotrends, accessed
July 14, 2025, https://www.macrotrends.net/global-metrics/countries/usa
/united-states/population.

214. Charles Marohn, "Driving Went Down. Fatalities Went Up. Here's Why," *Strong
Towns*, January 10, 2022, https://www.strongtowns.org/journal/2022/1/10
/driving-went-down-fatalities-went-up-heres-why.

215. Arialdi M. Miniño, Jiaquan Xu, Kenneth D. Kochanek, and Betzaida Tejada-Vera,
"Death in the United States, 2007," *NCHS* Data Brief no. 26 (December 2009), US
Department of Health and Human Services, Centers for Disease Control and Pre-
vention, National Center for Health Statistics.

216. One fatality every 4.2 million miles.

217. One fatality for every 3,400 people.

218. National Highway Traffic Safety Administration, "Fatality Analysis Reporting System: Fatal Crash Data Overview 2023, Washington, DC," US Department of Transportation, accessed July 14, 2025, https://www.nhtsa.gov
/research-data/fatality-analysis-reporting-system-fars.

219. Phrase coined in the "Highways or Dieways" ad campaign. See Rebecca
L. Evans, "'Highways or Dieways': Catchphrase or Truth in South Carolina?," November 18, 2002, http://thetandd.com/news/highways-or-dieways
-catch-phrase-or-truth-in-south-carolina/article_e6e1ec22-a584-510f-aa52
-5da88c20c9c8.html.

220. CDC, "Achievements in Public Health, 1900–1999, Motor-Vehicle Safety: A 20th
Century Public Health Achievement," *CDC-MMWR* 48, no. 18 (May 14, 1999):
369–74, http://www.cdc.gov/mmwr/preview/mmwrhtml/mm4818a1.htm.

221. Virginia Department of Transportation, "Traffic Counts," updated May 15,
2025, https://www.vdot.virginia.gov/doing-business/technical-guidance-and
-support/traffic-operations/traffic-counts/.

222. PlanIt Metro, "Data Download: Metrorail Ridership by Station by Month,
2010–2015," *PlanItMetro: Metro's Planning Blog*, March 24, 2016, https://planit-
metro.com/2016/03/24/data-download-metrorail-ridership-by-station
-by-month-2010-2015/.

223. Further along the transit line, the average weekday traffic is over 180,000 passengers per day, however. Shannon, "DC Mythbusting: Metro's Most Crowded,"

*We Love DC Blog*, November 10, 2009, http://www.welovedc.com/2009/11/10/dc-mythbusting-metros-most-crowded/.

224. For WMATA transit data, see "Data Download: May 2013–2014 Metrorail Ridership by Origin and Destination," *PlanItMetro: Metro's Planning Blog*, August 28, 2014, http://planitmetro.com/2014/08/28/data-download-may-2013-2014-metrorail-ridership-by-origin-and-destination/; for VDOT traffic data, see VDOT (2012) AADT by County, accessed September 18, 2025, https://www.vdot.virginia.gov/doing-business/technical-guidance-and-support/traffic-operations/traffic-counts/.

225. Richard A. Gould, *Archaeology and the Social History of Ships* (Cambridge University Press, 2000).

226. Todd Alexander Litman, with Eric Doherty, "Transportation Cost and Benefit Analysis: Techniques, Estimates, and Implications," *Victoria Transport Policy Institute*, January 2, 2009, https://vtpi.org/tca/tca01.pdf.

227. This and many more idiosyncrasies of America's parking priorities are presented and lampooned in Donald Shoup's *The High Cost of Free Parking* (Routledge, 2005).

228. Institute of Transportation Engineers, *Parking Generation*, 4th ed. (Institute of Transportation Engineers, 2010).

229. ULI Knowledge Finder, "Parking Policy Innovations in the United States," *ULI Knowledge Center Research Report*, May 12, 2021, https://knowledge.uli.org/en/reports/research-reports/innovations-in-parking-policy?q&sortBy=relevance&sortOrder=asc&page=1.

230. Alan Durning, "Apartment Blockers: Parking Rules Raise Your Rent," *Sightline Institute*, August 22, 2013, https://www.sightline.org/2013/08/22/apartment-blockers/.

231. The assumption is that people will not be able to reach your building at all unless you offer them parking. In much of America, this is true.

232. Litman, with Doherty, "Transportation Cost and Benefit Analysis."

233. Michael Manville, "Parking Requirements and Housing Development: Regulation and Reform in Los Angeles," *ACCESS Magazine* 44 (Spring 2014), accessed July 14, 2025, https://www.accessmagazine.org/spring-2014/parking-requirements-housing-development-regulation-reform-los-angeles/.

234. Richard Rothstein, *The Color of Law* (Liveright Publishing Corporation, 2018).

235. John M. MacDonald, Robert J. Stokes, Deborah A. Cohen, Aaron Kofner, and Greg

K. Ridgeway, "The Effect of Light Rail Transit on Body Mass Index and Physical Activity," *American Journal of Preventive Medicine* 39, no. 2 (August 2010): 105–12.

236. James R. Knickman and Emily K. Snell, "The 2030 Problem: Caring for Aging Baby Boomers," *Health Services Research* 37, no. 4 (August 2002): 849–84.

237. Robert A. Caro, *The Power Broker: Robert Moses and the Fall of New York* (Knopf, 1974).

238. Kelly Blue Book Editors, "Average Length of U.S. Vehicle Ownership Hit an All-Time High," February 23, 2012, http://www.kbb.com/car-news /all-the-latest/average-length-of-us-vehicle-ownership-hit-an-all_time-high/.

239. Matt Timmons, "Car Ownership Statistics in the U.S.," *Value Penguin Blog*, updated June 23, 2023, https://www.valuepenguin.com/auto-insurance/car -ownership-statistics.

240. Data on the price of a new car found in Stacy C. Davis and Robert G. Boundy, *Transportation Energy Data Book*, edition 40 (Oak Ridge National Laboratory, 2022); data on car ownership found in "Highway Statistics Summary to 1995."

241. Portland, OR, is considered a hub of bike culture, and its citywide bike share is 10 percent; William A. Pizer and Raymond Kopp, "Calculating the Costs of Environmental Regulation," *Resources for the Future* discussion paper, updated March 6, 2003, http://www.rff.org/Documents/RFF-DP-03-06.pdf.

242. American Automobile Association, "Your Driving Costs"; Aaron O'Neill, "Annual Gross Domestic Product and Real GDP in the United States from 1929 to 2022," *Statista*, accessed July 14, 2025, https://www.statista.com/statistics /1031678/gdp-and-real-gdp-united-states-1930-2019/.

243. "Highway Statistics Series," FHWA HF-210 financial data 1921–95 and "HF-10 series 1996–2010," updated June 2, 2025, http://www.fhwa.dot.gov/policy information/statistics.cfm.

244. National Conference of State Legislatures, "Special Registration Fees for Electric and Hybrid Vehicles," updated July 27, 2025, https://www.ncsl.org/transporta tion/special-registration-fees-for-electric-and-hybrid-vehicles.

245. Hong Yang, Mecit Cetin, and Qingyu Ma, "Guidelines for Using StreetLight Data for Planning Tasks," Virginia Transportation Research Council, March 2020, https://vtrc.virginia.gov/media/vtrc/vtrc-pdf/vtrc-pdf/20-R23.pdf.

246. Earl Swift, *The Big Roads: The Untold Story of the Engineers, Visionaries, and Trailblazers Who Created the American Superhighways* (Houghton Mifflin Harcourt, 2011).

247. "National Household Travel Survey, 1995, 2001, and 2009," accessed July 14, 2025, https://nhts.ornl.gov/.

248. The 2010 Census uses the American Community Survey, and I have not figured out how to get walking and biking trips distinct on that yet. Assuming that walkable, bikable, or transit-friendly places were likely to stay that way between 2000 and 2010, I used the 2000 census data on journey to work, even though it is over a decade old.

249. Notion comes from Steven Fleming, *Cycle Space: Architectural & Urban Design in the Age of the Bicycle* (NA010 Publishers, 2013), 25.

250. Assumed that the point coverage of city locations was near or within the CBD for cities.

251. US Census shapefiles of CBSAs found in "Smart Location Mapping," US EPA Office of Sustainable Communities, updated October 25, 2024, https://www.epa.gov/smartgrowth/smart-location-mapping.

252. Kevin Ramsey and Alexander Bell, Smart Location Database, updated October 25, 2024, https://www.epa.gov/smartgrowth/smart-location-mapping#SLD.

253. Note highway crossings and riparian routing.

254. Design block length is similar to walkable neighborhoods in historic cities, to allow multiple routes through urbanizing fabric.

255. Transit Line Capacity concept via Vukan Vuchic, *Transportation for Livable Cities* (Routledge, 2005), 12.

256. From publicly available data on energy consumption and passenger miles traveled for transit and traffic, I converted the per-mile energy consumption for walking and biking to British Thermal Units (BTU) per mile, converted BTU to Gallons of Gas Equivalents (GGE), and multiplied by the 2009 estimates of distance from the National Household Travel Survey. This provides the more familiar statistic of Passenger mpg (Pmpg)—the fuel efficiency of the vehicle divided by the average number of passengers carried by the vehicle.

257. BTU = British Thermal Units, a measure of energy. 1 BTU is equal to 1,055 Joules.

258. kW-h = KiloWatt-hours, a measure of energy. 1 kW-h is equal to 3,600,000 Joules.

259. Yuhan Huang, Nic C. Surawski, Bruce Organ, John L. Zhou, Oscar H. H. Tang, and Edward F. C. Chan, "Fuel Consumption and Emissions Performance Under Real Driving: Comparison Between Hybrid and Conventional Vehicles," *Science of the Total Environment*, vol. 659 (April 2019): 275–82.

260. Discussion of trends in American energy use at Alan Cunningham, "Drop It

Like It's Hot, Pt. 1," January 21, 2020, HeadWater Solutions Blog.com/2020/01/21 /drop-it-like-its-hot-pt-1/.

261. US Census Bureau, "Median and Average Square Feet of Floor Area in New Single-Family Houses Completed by Location," accessed September 19, 2025, https://www.census.gov/construction/chars/index.html.

262. Stefan Schmidhofer, Roland Leser, and Michael Ebert, "A Comparison Between the Structure in Elite Tennis and Kids Tennis on Scaled Courts (Tennis 10s)," *International Journal of Performance Analysis in Sport* 14, no. 3 (December 2014): 829–40.

263. International Basketball Federation, *Official Basketball Rules 2006,* March 31, 2006, https://lsssa.wordpress.com/wp-content/uploads/2013/11/fiba -basketball-rules-2006.pdf.

264. Grady L. Miller, "Baseball Field Layout and Construction," accessed July 14, 2025, https://www.baseball-almanac.com/stadium/baseball_field_construc tion.shtml.

265. S. Jhoanna Robledo, "Because We Wouldn't Trade a Patch of Grass for $528,783,552,000," *New York Magazine,* December 26, 2005, https://nymag .com/nymetro/news/reasonstoloveny/15362/.

266. Bureau of Transportation Statistics, National Transportation Statistics, "Personal Consumption Expenditures on Transportation by Subcategory," accessed September 19, 2025, https://web.archive.org/web/20170509183924/http://www .rita.dot.gov/bts/sites/rita.dot.gov.bts/files/publications/national_transportation _statistics/html/table_03_16.html; Erin Stepp, "*AAA* Reveals True Cost Of Vehicle Ownership," *AAA Newsroom,* August 23, 2017, accessed October 18, 2018, https://web.archive.org/web/20181013165719/https://newsroom.aaa .com/tag/driving-cost-per-mile/.

267. Bike & Walk Capital: Pedestrian & Bicycle Information Center, accessed September 19, 2025, archived at https://web.archive.org/web/20160311121707 /http://www.pedbikeinfo.org/data/library/details.cfm?id=4876; Maintenance: Sanderson Stewart, "Trail Asset Management Plan," Billings, Montana, May 24, 2011, https://www.billingschamber.com/media/Trail-Asset-Management -Plan-FINAL-June-2011.pdf; City of Greenville, SC (2011) Operations and Maintenance in Trails and Greenways Master Plan, accessed September 19, 2025, https://content.civicplus.com/api/assets/0d4d6ad5-c3fa-4081-b6ed-cdfffcb 63f59?scope=all; Joe McDermott, "Sidewalk Maintenance and Repair Plan," *City of Buenaventura*; "Milwaukee County Trails Network Plan, 2007," Milwaukee

County Department of Parks, Recreation, and Culture, https://county.mil
waukee.gov/ImageLibrary/Groups/cntyParks/Planning/trails/networkplan
/FinalTrailsNetworkPlan2007.pdf; Jeff Weinstein, Mike Sherrod, Noel Fearon,
and Derek Perez, "How Much Will That Trail Cost?," presentation to Trails and
Greenways Conference, May 11, 2007, https://www.yumpu.com/en/document
/view/23310782/how-much-will-that-trail-cost-california-state-parks#goo
gle_vignette; Sustainable Bethlehem, "Bethlehem, NY: Bicycle and Pedestrian
Program, Bicycle and Pedestrian Annual Operations Budget Recommenda-
tions," accessed July 11, 2025, http://www.townofbethlehem.org/Document
Center/View/3036; City of Wichita, KS, January 28, 2017, "Planning Level Cost
Estimator," https://web.archive.org/web/20170128161808/http://www.wichita
.gov/Government/Departments/Planning/PlanningDocument/Appendix%20
C%20-%20Planning%20Level%20Cost%20Estimator.pdf; Kimberley Turner,
"Where the Sidewalk Ends: Fixing Atlanta's Broken Footpaths," *Curbed: Atlanta,* August 1, 2014, http://atlanta.curbed.com/archives/2014/08/01/where
-the-sidewalk-ends-fixing-atlantas-broken-sidewalks.php; Operations: Ger-
ald Flores, "20 Interesting Sneaker Facts You Don't Remember," *Complex*, November 22, 2013, http://www.complex.com/sneakers/2013/11/20-interesting
-sneaker-facts-you-dont-remember/nike-freestyle-commercial; Sierra Club,
"Pedaling to Prosperity: Bicycling Will Save Americans $4.6 Billion in 2012,"
May 18, 2012, PDF available at https://unidosus.org/publications/641-pedaling
-to-prosperity-bicycling-will-save-americans-4-6-billion-in-2012/; Office of
Highway Policy Information "Highway Statistics 2007: Funding For Highways
and Disposition of Highway-User Revenues, All Units of Government, 2007,"
updated November 7, 2014, https://www.fhwa.dot.gov/policyinformation
/statistics/2007/hf10.cfm; "Government Transport Financial Statistics, 2007,"
Friday March 15, 2013, https://www.bts.gov/archive/publications/govern
ment_transportation_financial_statistics/2007/index; Bureau of Transportation Statistics, "Personal Consumption Expenditures on Transportation by
Subcategory," May 9, 2017, accessed September 19, 2025, https://web.archive
.org/web/20170509183924/http://www.rita.dot.gov/bts/sites/rita.dot.gov.bts
/files/publications/national_transportation_statistics/html/table_03_16.html;
Transit Capital: Federal Transit Administration (2024), "Current Capital Investment Grants Program and Expedited Project Delivery Pilot Program Projects," last updated September 9, 2025, http://www.fta.dot.gov/12304_14366

.html; Federal Transit Administration, "New Starts Project Profiles: An-nual Report on New Starts 2006," https://www.transit.dot.gov/sites/fta.dot .gov/files/docs/FY07_NEW_STARTS_REPORT_COMPLETE.pdf; Tran-sit Maintenance and Operations: American Public Transportation Associa-tion, "Public Transportation Fact Book," November 2015, https://www.apta .com/wp-content/uploads/Resources/resources/statistics/Documents/Fact book/2015-APTA-Fact-Book.pdf.

268. US Census Bureau Data Portal, accessed September 19, 2025, https://data.cen sus.gov/cedsci/.

269. John Puchner and Lewis Dijkstra, "Making Walking and Cycling Safer: Lessons from Europe," *Transportation Quarterly* 54, no. 3 (June 2000): 25–50; Ralph Bue-hler and John Pucher, "Cycling to Work in 90 Large American Cities: New Evi-dence on the Role of Bike Paths and Lanes," *Transportation* 39, no. 2 (July 2011): 409–32.

270. Stephen Miller, "New DOT Report: Protected Bike Lanes Improve Safety for Ev-eryone," *Streetsblog NYC*, September 5, 2014, https://nyc.streetsblog.org/2014 /09/05/new-dot-report-shows-protected-bike-lanes-improve-safety-for-every-body/; Nancy Scola, "What 5 Cities, 17K Cyclists, and 20K Cars Tell Us About Pro-tected Bike Lanes," Next City, accessed June 12, 2014, http://nextcity.org/daily/entry /us-city-bike-lanes-bike-safety.

271. Gary Pivo, and Jeffrey Fisher, "The Walkability Premium in Commercial Real Es-tate Investments," *Real Estate Economics* 39, no. 2 (Summer 2011): 185–219.

272. Emily Hamilton, "The Value of Walkability," *Market Urbanism*, October 17, 2013, https://marketurbanism.com/2013/10/17/the-value-of-walkability/.

273. Emily Badger, "The Simple Math that can Save Cities from Bankruptcy," *Bloomberg*, March 30, 2012, https://www.bloomberg.com/news/articles/2012-03-30/the -simple-math-that-can-save-cities-from-bankruptcy; Bruce Katz, "Smart Growth: The Future of the American Metropolis?," *Brookings Institution*, July 2002, https:// imfg.org/uploads/69/katz_casepaper58.pdf.

274. Joseph Minicozzi, "The Smart Math of Mixed-Use Development," *Planetizen: 25 Years in Planning*, January 23, 2012, http://www.planetizen.com/node/53922.

275. Energy Information Administration, "U.S. Energy Facts Explained," last updated July 15, 2024, https://www.eia.gov/energyexplained/us-energy-facts/.

276. Brenna Ellison, Linlin Fan, and Norbert L. W. Wilson, "Is it More Convenient to Waste? Trade-Offs Between Grocery Shopping and Waste Behaviors," *Agricultural*

*Economics* 53, issue S1 (November 2022): 75–89.

277. "Energy Flow Charts," Lawrence Livermore National Laboratories, accessed September 19, 2025, https://flowcharts.llnl.gov/. Data is based on DOE/EIA SEDS, accessed September 19, 2025, https://flowcharts.llnl.gov/commodities/energy.

278. "Energy Flow Charts," Lawrence Livermore National Laboratories.

279. Ranjitha Shivaram, Zheng Yang, and Rishee K. Jain, "Context-Aware Urban Energy Analytics (CUE-A): A Framework to Model Relationships Between Building Energy Use and Spatial Proximity of Urban Systems," *Sustainable Cities and Society* 72 (2021): 102978.

280. S. Faramawy, T. Zaki, and A. A.-E. Sakr, "Natural Gas Origin, Composition, and Processing: A Review," *Journal of Natural Gas Science and Engineering*, vol. 34 (August 2016): 34–54.

281. The "refrigerator problem" is a paradox of efficiency that observes that people offered a more efficient refrigerator will just get a bigger refrigerator, canceling out the efficiency savings of the improved technology. It is a warning to be aware if the demand for something is fixed or indefinite. Peter Huber, *Hard Green: Saving the Environment From the Environmentalists: A Conservative Manifesto* (Basic Books, 1999), 72.

282. Paula Manoela dos Santos, "The Eight Principles of the Sidewalk: Building More Active Cities," *The City Fix*, April 2, 2015, https://thecityfix.com/blog/the-eight-principles-of-the-sidewalk-building-more-active-cities-paula-santos/; Dom Nozzi, "Victor Dover's Five Basic Physical Features of Great Neighborhoods," November 18, 1998, http://walkablestreets.wordpress.com/1998/11/18/victor-dovers-five-basic-physical-features-of-great-neighborhoods/.

283. Richard D. Kahlenberg, *Excluded: How Snob Zoning, NIMBYism, and Class Bias Build the Walls We Don't See* (Hachette Book Group, 2023).

284. The League of American Bicyclists, "Long Term Bike Parking," https://bikeleague.org/sites/default/files/Long%20Term%20Parking%20Presentation_FINAL.pdf.

285. Sustainable America, "5 Steps to Building a Bike Friendly City," September 4, 2024, https://sustainableamerica.org/blog/five-steps-to-building-a-bike-friendly-city/.

286. Luminița Anica, "Vehicle-Ramming Attack—A New Modus Operandi of Terrorists," *International Scientific Conference Strategies XXI*, vol. 1 (2018): 67–79.

287. Rababe Saadaoui, Deborah Salon, Huê Tâm Jamme, Nicole Corcoran, and Jordyn Hitzeman, "Does Car Dependence Make People Unsatisfied with Life? Evidence from a U.S. National Survey," *Travel Behaviour and Society* 39 (April 2025): https://doi.org/10.1016/j.tbs.2024.100954.

288. Those rules were: 1) Give traffic as many lanes as required to reduce congestion for two hours a day; 2) Require off-street parking for all properties, each according to their need, and make that parking easy to find for drivers on the road; 3) Discourage or ban all vehicles from the road that cannot move at the ambient speed of traffic. Many states will not post a speed limit less than 25 mph. They cannot print a 1 on their speed limit signs; 4) Simplify driver decision-making while on the road.

289. "Loudoun County Bicycle and Pedestrian Mobility Master Plan," adopted October 20, 2003, 20, accessed September 19, 2025, https://www.loudoun.gov/DocumentCenter/View/1071/Bicycle--Pedestrian-Mobility-Master-Plan.

290. Brad Aaron, "A Streetside Chat with Jan Gehl," *Streetsblog NYC*, Nov 24, 2008, https://old.nyc.streetsblog.org/2008/11/24/streetfilms-a-streetside-chat-with-jan-gehl/.

291. Md. Kamruzzaman, Farjana Mostafiz Shatu, Julian Hine, and Gavin Turrell, "Commuting Mode Choice in Transit-Oriented Development: Disentangling the Effects of Competitive Neighbourhoods, Travel Attitudes, and Self-Selection," *Transport Policy* 42 (August 2015): 187–96.

292. "Quantitative Easing" (since 2008) and COVID stimulus (since 2020) injection of cash into the economy, combined with lockdowns and work from home has made many things more affordable for many Americans, but unattainable for Americans not fortunate enough to have cash.

293. "Smart Location Mapping," US EPA Office of Sustainable Communities.

294. "Energy Flow Charts," Lawrence Livermore National Laboratories.

295. Harald Sverdrup, Deniz Koca, and Kristin Valar Ragnarsdotti, "Investigating the Sustainability of the Global Silver Supply, Reserves, Stocks in Society and Market Price Using Different Approaches," *Resources, Conservation and Recycling* 83 (February 2014): 121–40.

296. Matti Sommarberg and S. J. Mäkinen, "Technopreneurial Characteristics Rising from the Ashes of Creative Destruction," paper presented at the *2018 Portland International Conference on Management of Engineering and Technology (PICMET)*, Institute of Electrical and Electronics Engineers, http://dx.doi.org/10.23919/PICMET.2018.8481846.

# Annotated Bibliography

Baltes, Michael R. "Factors Influencing Nondiscretionary Work Trips by Bicycle Determined from 1990 U.S. Census Metropolitan Statistical Area Data." *Transportation Research Record: Journal of the Transportation Research Board* 1538, no. 1 (January 1996): https://doi.org/10.1177/0361198196153800113.

A survey at the national scale of the metropolitan-level factors associated with more or less bike commuting. The progression of this and Zahran et al. from 2008 suggested to me that I should look at block groups and commuting at the station area, a geography that made the most sense for walking or biking.

Bernick, Michael, and Robert Cervero. *Transit Villages in the 21st Century*. McGraw Hill, 1996.

This was the first book I had read that introduced me to the term *transit-oriented development*, even though I grew up in a trolley suburb of Atlanta. I would soon visit places like Orangelington or Manhattan and read about places like Curitiba or Switzerland. Now readers can visit Curitiba and even Delhi on Google Street View.

Brand, Stewart. *How Buildings Learn*. Penguin, 1995.

This was the first book I read that lit up the notion that cities are a tapestry—a palimpsest—of the times in which, and since they were built. Money melts history.

Brandes Grartz, Roberta, and Norman Mintz. *Cities Back from the Edge—New Life for Downtown*. Wiley, 1998.

An optimistic book laying out a path to reviving downtowns, even before the influx of millennial young adults. Among its leading points, it suggests that cities are ill-served by oversized flagship projects, but do well when their development is incremental, prefiguring Strong Towns, and that transit should be better used, prefiguring the movement for transit-oriented development and this book. It also pointed out the primal importance of the real estate market to the success or failure of any downtown redevelopment. There have been at least two market crashes and a growing residential market distortion since this book was written, but this book remains a good roadmap for the fundamentals of stewarding a walkable, livable place back into business.

Cervero, Robert, and K. L. Wu. "Polycentrism, Commuting, and Residential Location in the San Francisco Bay Area." *Environment and Planning A: Economy and Space* 29, no. 5 (May 1997): https://doi.org/10.1068/a290865.

An important paper in the evolution of understanding metros. Though most of the centers considered were freeway exits, this made me think of transportation as the act of moving where we wanted to go, and how the aggregation of those desires into centers, edge cities, or transit-oriented developments between thousands of people could create places.

Condon, Patrick M. *Seven Rules for Sustainable Communities.* Island Press, 2010.

A transit-, transportation-, and land use-focused recommendation for sustainability. My only concern with it is that it was not fixed in place. One of the inspirations for the book about transit-oriented development as a default location for the strategies in this book.

Daniels, Tom. *When City and Country Collide: Managing Growth in the Metropolitan Fringe.* Island Press, 1999.

Sprawl happens at the rural fringe of American metropoles, one family farm lot at a time. This describes the cultural clash between rural populations focused on land and agricultural markets and new suburban homebuyers focused on traffic and job markets.

Duany, Andres, Elizabeth Plater-Zyberk, and Jeff Speck. *Suburban Nation: The Rise of Sprawl and the Decline of the American Dream.* North Point Press, 2000.

A foundational book for the New Urbanism movement, which has since spawned form-based design and the SmartCode. The seminal idea is that we have forgotten how to build at a human scale, both in buildings and in the arrangement of buildings on the American landscape. This is a clarion call to build at a human scale, starting with regulations for subdivisions, by expanding to zoning codes, county by county, revising their arrangement of development to serve walkers and bikers before traffic in downtowns, community centers, or transit-oriented developments.

Garreau, Joel. *Edge City: Life on the New Frontier by Joel Garreau*. Knopf Doubleday Publishing Group, 1991.

One of the books that started me thinking about the possibility of a multimodal
future in specific locations. Though this book was focused on the commercial and
residential potential of freeway exits, it could just as easily apply to transit stations.
Several of the edge cities mentioned in this book are now connected by rail transit
station to the rest of their metros.

Gehl, Jan. *Cities for People*. Island Press, 2010.

A challenge to the default notion that traffic is the inherent scale of cities. This
book describes in great detail the more durable status of people as walkers, not drivers. It explores how our use of cities as walkers shaped cities for millennia, and how
replacing walking as the default mode of transportation is dissolving cities.

Goddard, Stephen B. *Getting There: The Epic Struggle Between Road and Rail in the
American Century*. University of Chicago Press, 1994.

This is a historical description of the conflict between freight, intercity, and commuter rail networks, which were the dominant modes of transportation in the nineteenth century, and the emerging road networks and their traffic, which were dominant by the twentieth century. Roads were filled with individual passenger and
freight vehicles, and rail networks were already well established, offering many
more opportunities for tragic collisions than before. This book opened my eyes to
the economic weight of railway monopolies—as most cities had a transit monopoly—and why traffic was seen as such a democratizing revolution.

Hawken, Paul. *The Ecology of Commerce: A Declaration of Sustainability*. Harper-
Collins, 1993.

The notion that sustainable solutions provide value for a wide sector of the population and can be valued and be market assets in the population to foster envi-
ronmentally sensible outcomes with minimal influence from government power.

Hayden, Dolores. *The Power of Place: Urban Landscapes as Public History*. MIT
Press, 1997.

One of my introductions to the idea that cities and places are ephemeral and dynamic, as indefinite as their infrastructure or economies might be.

Huber, Peter. *Hard Green: Saving the Environment from the Environmentalists—A Conservative Manifesto*. Basic Books, 1999.

A reminder that environmental and social goals are more durably achieved through market and social forces than top-down policies and guilt.

Hudson, William. "Myths and Milestones in Bicycle Evolution." Accessed July 15, 2025. https://jimlangley.net/ride/bicyclehistorywh.html.

Comprehensive history of the development of bicycle technology. There are similar resources, like http://crazyguyonabike.com, that are sadly no longer available.

International Bicycle Fund. "Bicycle History." Accessed July 15, 2025. http://www.ibike.org/library/history-timeline.htm.

An informative history of the bicycle from its beginning as custom craft and engineering, through fads in the nineteenth century, to economic alternatives to walking and the use of the horse by the 1890s, just in time for traffic to take over the streets in America.

Jackson, Kenneth T. *Crabgrass Frontier: The Suburbanization of the United States*. Oxford University Press, 1985.

An arresting narrative of the appeals and cost of America's breakneck suburbanization, including a deeper exploration of the history of the suburb as apart from the city, dating back to ancient times. Commuter rail and the trolley were once responsible for sprawl in their own time.

Jacobs, Jane. *The Death and Life of Great American Cities*. Random House, 1961.

This is a love letter to the last bastion of walkable urbanism in the US: New York City. The book offers astute understandings of the balance between walkers, building types, street level uses, and how traffic distorts and destroys those relationships.

Jones, David W. *Mass Motorization and Mass Transit*. Indiana University Press, 2008.

The most dog-eared book in my collection, as this book is a font of quantitative information on the transformation of America from the leading transit user by the end of the nineteenth century to the leading traffic user by the end of the twentieth. This lays out, in quantitative terms, why traffic was so compelling and useful to Americans even as a luxury by the 1910s, and as a competitor for street space, especially when the used car market bloomed in the form of jitneys for hire.

Kay, Jane Holtz. *Asphalt Nation: How the Automobile Took Over America and How We Can Take It Back*. University of California Press, 1997.
This book, like several others in this bibliography condemns sprawl and spends the first three quarters of the text doing that in detail, but offers a solid last quarter on solutions. Reading this was about the time I decided to write this book and not follow the same plot, as that apparently was not working to improve transportation or land use.

Kunstler, James Howard. *The Geography of Nowhere: The Rise and Decline of America's Man-Made Landscape*. Free Press, 1994.
I saw Kunstler when he gave a talk at Clemson's school of architecture, where I was studying GIS as a skill for landscape ecology. He solidified for me the urgency of why landscape ecology alone was not going to do much good in the world, and that the better course was in planning and the sausage-making of governance policy and civil engineering. This book itself is a jeremiad against the suburban norm, and the manifest unworkability and unaffordability of a landscape built solely for traffic.

Lang, Robert. *Edgeless Cities: Exploring the Elusive Metropolis*. Brookings Institution Press, 2003.
A book that showed me that the transportation modes we use commit us to certain urban forms, notwithstanding design. This highlighted the urgency of building a case for sustainable transportation modes in an America that had seemed destined for traffic for my grandparents' entire lives.

Lay, M. G. *Ways of the World: A History of the World's Roads and of the Vehicles That Used Them*. Bedford/St. Martin's, 1992.
A deep, eye-opening history of the ways that we have expanded pathways into trackways, railways, and roadways over the last three or four millennia.

Leavitt, Helen. *Superhighway–Superhoax*. Doubleday & Company, 1970.
This strident warning against the superhighway system, still under construction, was not so much about sprawl as the safety, time, and infrastructure costs of

our commitment to the interstate highway networks still being built in and on our cities at the time of publication.

Leinberger, Christopher B. *The Option of Urbanism: Investing in a New American Dream*. Island Press, 2009.

A catalog and profile of walkable edge cities within metropolitan sprawl, with the context of America's suburbanization. Almost all the examples are upmarket and desirable loci within prosperous cities, which points to the need for a growing regional economy and real estate financing for walkable urbanism to be built. This was one of the reasons I did not recommend anything more than upzoning to minimally walkable intensities, as walkable urbanism was still seen in 2009 as a luxury good to be driven to or purchased.

Levine, Jonathan. *Zoned Out: Regulation, Markets, and Choices in Transportation and Metropolitan Land Use*. RFP Press, 2006.

This book was my introduction to the market distortions of zoning and how it forced sprawl on metropolitan landscapes—one of the seeds for this book, which asked: how can we grow inward when the default option is to grow outward?

Levinson, Marc. *The Box: How the Shipping Container Made the World Smaller and the World Economy Bigger*. Princeton University Press, 2016.

While adding Jacobs to this bibliography, I recalled her praise of diners populated by all walks of life as an equalizing American "third place," including stevedores getting meals between shipments at the Manhattan docks on the Lower West Side. This book explains why that was an ephemeral mixture, and why there's a huge pile of multicolored, multicultural containers in residence near the IKEA in Newark.

Lewyn, Michael. *A Libertarian Smart Growth Agenda: How to Limit Sprawl Without Limiting Property Rights*. Lambert Academic Publishing, 2012.

A great reminder that the best way to get things done is through enabling markets for walkable urbanism, not by regulating its conditions. Sprawl is largely the result of government regulations of land. What would be the consequences of a looser approach to regulation?

Lucy, William H., and David I. Phillips. *Tomorrow's Cities, Tomorrow's Suburbs*. APA Planners Press, 2006.

A forward-looking book about possible directions in the development of a city. Like every forecast book, it contains the rumblings of current issues with different names, as well as past concerns we no longer worry about, while missing other powerful trends entirely. The most important lesson for me was that cities are malleable and changing, and are never doomed for long, as long as they still have opportunity and their people have agency.

Mapes, Jeff. *Pedaling Revolution: How Cyclists Are Changing American Cities*. Oregon State University Press, 2009.

A hopeful book profiling the rise of biking safety culture and place in American cities, including Davis, CA, Portland, OR, and the still rising bike lanes in New York City. Though the book models the bikeable city on Amsterdam, each American city considered is still not nearly as bike-oriented as Amsterdam, Freiburg, or even Tokyo—yet.

Marohn Jr., Charles. *Confessions of a Recovering Engineer: Transportation for a Strong Town*. Wiley, 2021.

The author has gone through a great transformation, in that he actually has a PE and is a civil engineer who became disillusioned with the market and safety hazards of our commitment to traffic as our sole transportation mode. This makes this book an invaluable resource for changing the way we do things, because he knows how the sausage is made. I was skeptical of traffic dependence since growing up biking in Atlanta. Marohn came to realize the costs of traffic dependence as a result of working for it

Martinson, Tom. *American Dreamscape: The Pursuit of Happiness in Post-War Suburbia*. Carroll & Graf, 2000.

The first book I read that did not see sprawl as a categorical evil, but an invitation to diversity and opportunity. Foundational to my consideration of proposing a way that would enable walking, biking, and transit to have a place alongside traffic, not in competition with it.

McCann, Barbara. *Completing Our Streets: The Transition to Safe and Inclusive Transportation Networks*. Island Press, 2013.

An adopted recommendation for true multimodal transportation within the right of way. The book stays in the scope of departments of transportation and therefore has a place in the design toolbox of many state and local DOTs.

McCourt, Frank. *'Tis*. Scribner, 1999.

The only memoir in this list. There was a scene where the impoverished family was tearing apart walls in their home for firewood. The book provides a striking illustration of why old homes are so well-built and scarce: survivor bias, not any lost ancient craft in the halcyon past.

McShane, Clay. *Down the Asphalt Path: The Automobile and the American City*. Columbia University Press, 1994.

A great description of the infrastructural and behavioral changes from the horse-drawn and trolley-dominated 1890s to the traffic-dominated 1930s, including a description of the initial appeal of traffic—the call of bucolic suburbia, relaxed gender roles, and unprecedented freedom of automobilism—and the urgent development of traffic safety.

Mumford, Lewis. *The Highway and the City*. Praeger, 1981.

A classic takedown of the decades-long zeal for highway maximalization in America, contrasting European and American negotiations between traffic and cities.

National Transportation Atlas Database: 2022. https://www.bts.gov/ntad.

Along with TranStats, this has been a great source for geography related to transit and traffic at the national scale.

Newman, Peter, Timothy Beatley, and Heather Boyer. *Resilient Cities: Responding to Peak Oil and Climate Change*. Island Press, 2009.

This book reminds me that peak oil is an imperfect driver for controlling urbanism or sprawl. As the cost of fuel rises, consumers try to find ways to use less, but producers are in a better position to produce more, America is now the leading exporter and producer of fossil fuels, as hydraulic fracking has become safe and reliable. Oil drilling itself only developed between 1850 and 1870, after all.

Newman, Peter, and Jeffrey Kenworthy. "Urban Design to Reduce Automobile Dependence." *Opolis* 2, no. 1 (Winter 2006): 35–52.
In writing this book, I searched for a standard density for walkability. This paper provided it, in 14 HU+Jobs/acre. This is a metric for overall metros, but it is astonishing how many neighborhoods and census block groups fail to have this basic intensity in America. Of course, we've been building solely for traffic for a century now. Time for something new.

Norquist, John O. *The Wealth of Cities: Revitalizing the Centers of American Life.* Perseus Books Group, 1998.
There is nothing immutable about cities or inherently superior about their form, they are an artifact of building and transportation technology, economies, and societies. This is an inspirational catalog of the advantages of cities as gathering places for culture and resilience.

Oldenburg, Ray. *The Great Good Place: Cafés, Coffee Shops, Bookstores, Bars, Hair Salons and Other Hangouts at the Heart of a Community.* Paragon House, 1989.
In order to have a sense of community where we are living, working, shopping, or even playing, community gathering places where all generations and occupations are welcome are vital to the stability and identity of those places. These "third places" are fragile if everyone needs to drive to and from them.

Portney, Kent E. *Taking Sustainable Cities Seriously: Economic Development, The Environment and Quality of Life in American Cities.* MIT Press, 2013.
This covers the broad application of sustainability to large- and medium-sized cities in America, which includes economic, social, and environmental stewardship. Transportation choice and land use availability for different modes touches on each of these, but as this book shows, sustainability, like economics, is a multipronged effort.

Pucher, John, and Ralph Buehler. *City Cycling*. MIT Press, 2012.

This book describes the state of the art in cycling safety practices and infrastructure recommendations, including detailed discussion of small- and large-city practices, as well as practices for women and children to enhance the usefulness of cycling as a viable transportation mode among cities built out to serve traffic.

Ross, Benjamin. *Dead End: Suburban Sprawl and the Rebirth of American Urbanism*. Oxford University Press, 2014.

This book does a better job than most at laying out how our transportation, zoning, and subdivision policies destined America for sprawl development, and how to shift it. The author's recommendation to build more transit rang hollow for me, because we don't even use most of the transit we've got nearly as much as we could. That was about when I began conceiving of this book.

Rowe, Adam. *The Bulldozer in the Countryside: Suburban Sprawl and the Rise of American Environmentalism*. Cambridge University Press, 2001.

A history of the third pulse of roadbuilding and suburbanization, and the first one to exceed the bounds of commuter rail transit. The scale of building—with the dawn of the interstates and enabling of places like Levittown—gave birth to a new Environmentalism in America, birthing many of today's federal environmental laws like the Clean Air Act and Clean Water Act.

Rybczynski, Witold. *Last Harvest: From Cornfield to New Town*. Scribner, 2007.

An intimate and detailed account of the conversion of a farm into suburban housing, showing the politics, personalities, and risks in real estate development, by one of the best writers I've ever read.

Sadik-Khan, Janette, and Seth Solomonow. *Street Fight: Handbook for an Urban Revolution*. Penguin, 2017.

A great play-by-play of the New York streets commissioner undoing the work of the former parks commissioner, Robert Moses. New York's quality of life increased greatly during the Bloomberg administration, partly because of Sadik-Khan's work allowing walkers back onto its streets.

Safdie, Moshe, and Wendy Kohn. *The City After the Automobile: An Architect's*

*Vision*. Taylor & Francis; Westview Press, 1998.

 A visionary consideration of life after the automobile, including a consideration of cybernetic disconnection from spatial constraints.

Shoup, Donald. *The High Cost of Free Parking*. Routledge, 2005.

 A systematic repudiation of the "science" of minimum parking requirements that are imposed on all landowners in the US, and an interrogation of the assumption that building too much parking for traffic will do anything but increase the demand for traffic.

Speck, Jeff. *Walkable City: How Downtown Can Save America, One Step at a Time*. Farrar, Straus, and Giroux, 2012.

 A comprehensive description of the benefits, practices, and fragility of walkable urbanism. The subtitle was a source of despair for me, as Americans should have access to walkability in many more places than just downtown. Let the walkable city have equal standing with the drivable metropolis.

Sucher, David. *City Comforts – How to Build an Urban Village*. City Comforts Press, 1995.

 This is one of the books about urbanism and the convenience of walkability and accessibility that told me we were not lacking for design solutions for walkable urbanism, but needed to allow it in America—very much at the neighborhood scale.

Swift, Earl. *The Big Roads—The Untold Story of the Engineers, Visionaries, and Trailblazers Who Created the American Superhighways*. Mariner Books, 2011.

 The story of the development of the federal and state highway departments from farm roads and the concern of agriculture in the 1890s to the modern hierarchical traffic system administered mostly by the states with the guidance of the federal highway department. An engaging read that highlights what Americans take for granted in roadway design.

Thaler, Richard, and Cass Susstein. *Nudge: Improving Decisions About Health, Wealth, and Happiness*. Penguin, 2008.

 A book in tension between libertarian and economic principles, it reminds us that there are much better ways than edicts to change human behavior, such as controlling attention. The book lurches over to marketing but nevertheless makes a great case for using a softer touch than policy. This is another reason I was not overly prescriptive in my book, after seeing the results of decades of "urban renewal" attempts.

Thompson, Joe. The Cable Car Page. Accessed July 15, 2025. https://www .cable-car-guy.com/.
There is brief mention of cable cars in this book, but for a couple of decades they were in use as the primary means of mass transportation in about twenty cities in the US. This site provides a great history of that period, with plenty of mentions of conditions between 1870 and 1890.

TranStats. Bureau of Transportation Statistics 2025. Accessed July 15, 2025. https:// www.transtats.bts.gov/.
I started working on this book in 2011, and this was a font of information for my early concepts. This page has changed massively, but is still a great resource for primary statistics in a fairly timely platform.

Vanderbilt, Tom. *Traffic: Why We Drive the Way We Do*. Vintage, 2009.
A great profile of the behavioral act of driving, in relation to drivers, vehicles, and the people around drivers, from the sidewalk, suburbs, or cities.

Vuchic, Vukan. *Transportation for Livable Cities*. Routledge, 1999.
A technical and detailed exposé of the perverse incentives of traffic and a pathway to a multimodal future. Vuchic shows we already know how to plan transportation that works for cities. Much of the reason I wrote this book was to point out where to plan cities to work for transportation.

Walker, Jarrett. *Human Transit: How Clear Thinking About Public Transit Can Enrich Our Communities and Our Lives*. Island Press, 2012.
A profile of the multiple ways that transit could be and should be designed to provide the best service to passengers, origins, destinations, places, and cities.

Zahran, Sammy, Samuel D. Brody, Praveen Maghelal, Andrew Prelog, and Michael Lacy. "Cycling and Walking: Explaining the Spatial Distribution of Healthy

Modes of Transportation in the United States." *Transportation Research Part D: Transportation and Environment* 13, no. 7 (October 2008): 413–83.

A national survey of the county-level factors associated with walking and biking. Looking at wide areas in fine detail has always appealed to me and provided the inspiration for many of the maps in this book.

# Index

property, 151–52; pavement and, 32; right
of way, 22, 58, 64; tax, 136, *136*; values,
97, 136
proximity: of building, 139; housing, 91–94,
95, *96*; job, 91–94, 95, *96*, 111; to shopping, 61, *62*, 91–94, *96*; to transit stop,
91–94, 95, *96*; walking, 91–94, 95, *96*

quality of life: benefits, 143–46; traffic and,
39–45, *45*, 172n187; transit station and,
143–46

rail, 27, 32, 76; development of, 22, *23*, *37*;
passenger trains, 152–53; stations, 143,
151. *See also* transit, rail
railroad, *37*; elevated, 26–27; property
right of way by, 22, 58, 64; robber barons, 27–28
rainfall, 47
Randel Farm Map (1820), 160n44
real estate, 116; market and zoning for, 146–
47; premium, 135–37, *136*; tax return
per acre for, 136, *136*. *See also* building;
housing; land
refrigerator problem, 141, 183n281
right of way, 63–64, 150; intersection, 172n187; law, death and, 54–58,
174n204, 174n208, 175n210; property,
22, 58, 64; standards, 44–45
road, 152; maintenance, *47*, 72, *73*; size,
peak demand and, 84
road network, 1, 20, *21*, 41–44; agriculture
and, 48–50, *50*; benefits of, 39–45, *45*,
48–50, *49*, 172n187; clear zone of, 60;
dashes on, 62–63; design of, 9–10, *37*;

engineering of, 51–53, *52*, *53*, 173n195;
fatalities in, 51–53, *52*, *53*; financing of,
33–34, 45; pavement of, 31–33, 43, 64,
168n144, 168n146; peak loads of, 64–65;
water in, 47, 48, 61
rubber, synthetic, 32

Sadik-Khan, Janette, 69
safety: biking, 122–23, *123*, 127; highway
acts, 60; stopping distance and, 62–
63, *63*; traffic, 44–45, *45*, 122–23, *123*,
127; traffic congestion and, 53, *53*; transit, 122–23, *123*, 127; walking, 122–23,
*123*, 127
San Francisco, 24, 84, 163n77
schools, 108–9
scooter, 29
sewer management, 18–20; in cities,
157n22; horses and, 23–24
shopping, 12; proximity to, 61, *62*, 91–94,
*96*; strip mall, 144
sidewalk, 95, 103, *105*; maintenance of, 120;
users of, 116–17, *118*; width of, 116–17, *117*
signs, 7–8
Smart Growth, 12, 69
Smith, Arthur, 57, 174n208
South America, 15–16
space: for biking, 115–19, *117*, *118*, *119*; lane,
142, 184n288; per passenger, 116, *117*; for
traffic, 115–19, *117*, *118*, *119*; traffic cost
for, *59*, 60–64, *62*, *63*, 176n223, 177n227,
177n231; for transit, 115–19, *117*, *118*, *119*;
usage, 130, *131*, 132–33; for walking, 115–
19, *117*, *118*, *119*
Spain, 16

# About the Author

Alan Cunningham, MS, MCP, PhD, LEED-AP, AICP, is a transportation planner, landscape ecologist, and watershed conservationist who has worked with civil engineering as well as state and metro community planning efforts, including award-winning and innovative work in Philadelphia, Atlanta, and Washington.

www.ingramcontent.com/pod-product-compliance
Lightning Source LLC
Chambersburg PA
CBHW070909270326
41927CB00011B/2500